# PRINCIPLES AND TECHNIQUES OF
# BIOCHEMISTRY AND MOLECULAR BIOLOGY

# PRINCIPLES AND TECHNIQUES OF
# BIOCHEMISTRY AND MOLECULAR BIOLOGY

**Evelyn Moss**
**Weim Huge**

Kruger Brentt
Publishers
2025

Kruger Brentt Publishers UK. LTD.
Company Number 9728962

Regd. Office: 68 St Margarets Road, Edgware, Middlesex HA8 9UU



ISBN: 978-1-78715-447-6

*Disclaimer:*

*Every possible effort has been made to ensure that the information contained in this book is accurate at the time of going to press, and the publisher and author cannot accept responsibility for any errors or omissions, however caused. No responsibility for loss or damage occasioned to any person acting, or refraining from action, as a result of the material in this publication can be accepted by the editor, the publisher or the author. The Publisher is not associated with any product or vendor mentioned in the book. The contents of this work are intended to further general scientific research, understanding and discussion only. Readers should consult with a specialist where appropriate.*

*Every effort has been made to trace the owners of copyright material used in this book, if any. The author and the publisher will be grateful for any omission brought to their notice for acknowledgement in the future editions of the book.*



For information on all our publications visit our website at http://krugerbrentt.com/

# PREFACE

The principles and techniques of biochemistry and molecular biology form the foundation of modern biological research, advancing our understanding of cellular functions, molecular interactions, and genetic mechanisms. These fields intersect at the molecular level, studying the intricate structures and reactions that make life possible.

Biochemistry, the study of chemical processes within and related to living organisms, investigates biomolecules like proteins, nucleic acids, carbohydrates, and lipids, emphasizing their structure, function, and interactions. Molecular biology, on the other hand, focuses on the molecular underpinnings of genetic information transfer, including DNA replication, RNA transcription, and protein synthesis. Together, these disciplines explore how cells operate, respond to stimuli, and maintain their physiological states.

Key techniques in biochemistry include spectrophotometry, chromatography, electrophoresis, and mass spectrometry, which allow for the isolation, identification, and quantification of biomolecules. Spectrophotometry, for example, measures the absorbance or transmission of light by a sample, often used to determine protein or nucleic acid concentration.

Chromatography enables the separation of molecules based on differences in properties like size or charge, essential for purifying complex mixtures. Electrophoresis is widely used to separate DNA, RNA, or proteins based on size and charge, providing insights into genetic material and protein structure. Mass spectrometry, which analyzes molecules by their mass-to-charge ratio, is invaluable for identifying biomolecular composition and structural information.

In molecular biology, DNA manipulation techniques, such as polymerase chain reaction (PCR), gel electrophoresis, and sequencing, have revolutionized genetic research. PCR amplifies specific DNA segments, making it possible to study genes in detail, even from small samples. Sequencing technologies, including next-generation sequencing (NGS), allow for rapid and comprehensive analysis of entire genomes, helping to identify genetic variations and mutations associated with diseases. Recombinant DNA technology is another powerful tool, enabling the insertion of specific genes into host cells to produce desired proteins, a process critical in developing therapeutic proteins and vaccines.

Overall, the principles and techniques of biochemistry and molecular biology are indispensable in modern science, enabling researchers to dissect cellular processes and genetic pathways at unprecedented levels. These fields have paved the way for significant advancements in medicine, biotechnology, and pharmacology, including drug development, gene therapy, and diagnostic tools. As research technologies continue to evolve, biochemistry and molecular biology will remain central to addressing complex biological questions and solving real-world health challenges.

**Evelyn Moss**

**Weim Huge**

# CONTENTS

# CHAPTER 1

# BIOCHEMICAL AND MOLECULAR BIOLOGICAL METHODS IN LIFE SCIENCES STUDIES

Biochemical and molecular biological methods are central to advancing our understanding of life sciences, as they provide insights into the molecular mechanisms underpinning cellular function, disease processes, genetic inheritance, and more. These methods encompass a diverse array of techniques designed to analyze and manipulate molecules like DNA, RNA, proteins, and metabolites within biological systems. By employing biochemical and molecular approaches, scientists can unravel the complexity of biological processes at the most fundamental level, leading to groundbreaking discoveries in fields such as medicine, genetics, microbiology, and biotechnology.

## 1.1 DNA AND RNA EXTRACTION

DNA and RNA extraction is a fundamental process in molecular biology that involves isolating genetic material from cells or tissues. This step is essential for various downstream applications, such as PCR, sequencing, and gene expression studies. During extraction, cells are lysed to release their contents, and then the DNA or RNA is purified through a series of chemical treatments and centrifugation steps. Several commercial kits are now available, simplifying the process and improving the yield and purity of extracted genetic material. The extracted DNA or RNA provides the basis for analyzing genetic sequences, identifying mutations, and studying gene expression patterns in various organisms, from bacteria to humans.

## 1.2 POLYMERASE CHAIN REACTION (PCR) AND REAL-TIME PCR

The Polymerase Chain Reaction (PCR) is a pivotal technique used to amplify small quantities of DNA to produce millions of copies. This process enables researchers to study specific DNA sequences in detail, even if they are present in minute amounts in a sample. PCR is essential for applications such as genetic testing, cloning, forensic science, and diagnosing infectious diseases. Real-time PCR, also known as quantitative PCR (qPCR), is an advanced version that measures DNA amplification in real time, allowing researchers to quantify gene expression levels accurately. PCR and qPCR have revolutionized molecular biology by making it possible to analyze genetic material rapidly and precisely, enhancing our ability to understand gene function, detect pathogens, and study genetic disorders.

## 1.3 GEL ELECTROPHORESIS

Gel electrophoresis is a technique used to separate DNA, RNA, or protein molecules based on their size and charge. This method involves loading samples onto a gel matrix and applying an electric current, causing the molecules to migrate through the gel. Smaller molecules move faster and farther through the gel than larger ones, resulting in a banding pattern that can be visualized with staining techniques. Gel electrophoresis is commonly used to verify the success of DNA amplification, analyze restriction enzyme digests, and study protein size and purity. This technique is essential for confirming experimental results and preparing samples for further analysis, such as sequencing or mass spectrometry.

## 1.4 DNA SEQUENCING

DNA sequencing is a method used to determine the precise order of nucleotides within a DNA molecule. With advancements in sequencing technologies, such as next-generation sequencing (NGS), researchers can sequence entire genomes quickly and cost-effectively. DNA sequencing has transformed life sciences by enabling the identification of genetic variations, such as single nucleotide polymorphisms (SNPs) and mutations associated with diseases. Applications of DNA sequencing range from human genome research and personalized medicine to evolutionary biology and microbial ecology. Sequencing allows researchers to decode genetic information, providing insights into gene function, evolutionary relationships, and the genetic basis of diseases.

## 1.5 WESTERN BLOTTING AND PROTEIN ANALYSIS

Western blotting is a widely used technique in protein analysis that detects specific proteins in a complex sample. This method involves separating proteins by gel electrophoresis, transferring them onto a membrane, and probing them with antibodies specific to the target protein. Western blotting is crucial for studying protein expression levels, post-translational modifications, and cellular signaling pathways. Protein analysis, more broadly, includes techniques like enzyme-linked immunosorbent assay (ELISA), mass spectrometry, and chromatography. These methods allow researchers to study the structure, function, and interactions of proteins, which are essential for understanding cellular processes, disease mechanisms, and drug targets.

## 1.6 ENZYME-LINKED IMMUNOSORBENT ASSAY (ELISA)

ELISA is a sensitive technique used to detect and quantify specific proteins, antibodies, hormones, or other molecules in a sample. This method relies on antigen-antibody interactions and involves coating a plate with an antigen or antibody, adding the sample, and then using an enzyme-linked antibody to produce a measurable color change. ELISA is commonly used in diagnostics, such as detecting infections, measuring immune responses, and monitoring disease progression. In research, ELISA provides a powerful tool for studying protein expression, identifying biomarkers, and analyzing immune responses in various biological systems.

## 1.7 RECOMBINANT DNA TECHNOLOGY AND GENE CLONING

Recombinant DNA technology involves combining DNA from different sources to create new genetic combinations. This process enables scientists to clone genes, express proteins, and introduce genetic modifications into organisms. Gene cloning, a core aspect of recombinant DNA technology, involves inserting a target gene into a plasmid vector, which is then introduced into a host organism for replication. This method is fundamental to biotechnology, as it allows for the production of therapeutic proteins, development of genetically modified organisms (GMOs), and study of gene function. Recombinant DNA technology has enabled advances in gene therapy, agricultural biotechnology, and drug development, making it a cornerstone of modern molecular biology.

## 1.8 CRISPR-CAS9 AND GENOME EDITING

CRISPR-Cas9 is a revolutionary genome editing technology that allows precise alterations to be made in an organism's DNA. This method uses a guide RNA to target a specific DNA sequence, while the Cas9 enzyme cuts the DNA at the designated site. Researchers can then introduce genetic changes by inserting, deleting, or replacing DNA at the cut site. CRISPR has transformed genetics and biomedical research by enabling the study of gene function, development of disease models, and exploration of potential therapeutic interventions. The technology holds immense promise for curing genetic disorders, developing new treatments for cancer, and engineering organisms for agricultural and industrial applications.

## 1.9 FLOW CYTOMETRY

Flow cytometry is a technique used to analyze the physical and chemical characteristics of cells or particles in suspension. This method involves passing cells through a laser beam, where they scatter light and emit fluorescence based on labeled antibodies. Flow cytometry is invaluable for studying immune cells, analyzing cell populations, and detecting rare cell types in a sample. Applications of flow cytometry include immunology, cancer research, and stem cell studies, where it enables researchers to sort cells, assess cell health, and analyze cell surface markers. This technology provides detailed insights into cellular heterogeneity and function, facilitating advancements in diagnostic and therapeutic research.

## 1.10 PROTEOMICS AND MASS SPECTROMETRY

Proteomics is the large-scale study of proteins, including their structure, function, and interactions. Mass spectrometry is a key tool in proteomics, enabling the identification and quantification of proteins in complex mixtures. This technique measures the mass-to-charge ratio of ions to determine the composition and structure of proteins. Proteomics and mass spectrometry are used to study protein pathways, identify disease biomarkers, and analyze post-translational modifications. These approaches provide a comprehensive view of the protein landscape within a cell or organism, offering insights into cellular processes and disease mechanisms at the molecular level.

## 1.11 CONCLUSION

Biochemical and molecular biological methods are essential tools in life sciences, enabling researchers to explore the fundamental mechanisms of life and disease at the molecular level. Techniques such as DNA and RNA extraction, PCR, sequencing, Western blotting, ELISA, and CRISPR-Cas9 have transformed our ability to analyze and manipulate biological systems. These methods have facilitated advances in medical diagnostics, drug development, agriculture, and environmental science, and they continue to drive innovations across a range of scientific fields. As technology progresses, biochemical and molecular methods will undoubtedly evolve, providing deeper insights and unlocking new possibilities for understanding and improving life on Earth.

# CHAPTER 2
# BASIC PRINCIPLES

Biochemistry and molecular biology are foundational sciences in the study of life, focusing on the molecular processes that drive cellular function and organismal development.

These fields combine the principles of chemistry, biology, and physics to explore the structure, function, and interactions of biomolecules like proteins, nucleic acids, lipids, and carbohydrates. This approach provides crucial insights into the molecular mechanisms that underpin cellular processes, genetics, and biochemical pathways, paving the way for advancements in medicine, genetics, biotechnology, and environmental science.

## 2.1 THE PRINCIPLE OF STRUCTURE AND FUNCTION

One of the fundamental principles in biochemistry and molecular biology is that the structure of a molecule determines its function. This concept is especially important in the study of proteins, where the three-dimensional folding of amino acid chains creates specific shapes that enable proteins to interact with other molecules and catalyze biochemical reactions.

The relationship between structure and function also applies to nucleic acids like DNA and RNA, where the double-helix structure of DNA allows for its replication and encoding of genetic information. By studying the structures of biomolecules, scientists can infer their functions and roles in biological processes, which is essential for drug design, genetic engineering, and enzyme research.

## 2.2 THE CENTRAL DOGMA OF MOLECULAR BIOLOGY

The central dogma of molecular biology is a foundational concept describing the flow of genetic information within a cell. It states that genetic information is stored in DNA, transcribed into RNA, and then translated into proteins. This DNA-RNA-protein flow is a central mechanism by which genetic instructions are executed to build and maintain cells. DNA serves as the blueprint, RNA as the messenger, and proteins as the functional molecules that carry out cellular processes. This principle underpins a vast array of molecular biology techniques, such as gene expression analysis, RNA sequencing, and protein synthesis studies, all of which have applications in genetics, medicine, and biotechnology.

## 2.3 ENZYME KINETICS AND CATALYSIS

Enzymes are biological catalysts that accelerate chemical reactions within the cell. Understanding enzyme kinetics—the study of the rates at which enzyme-catalyzed reactions occur—is crucial in biochemistry. Enzyme kinetics helps scientists determine how fast reactions proceed, the factors that affect enzyme activity, and how enzymes interact with substrates. This knowledge is vital for drug development, as many drugs function by inhibiting enzyme activity in disease pathways. Techniques such as spectrophotometry and chromatography are often used to measure enzyme activity, providing insights into the biochemical pathways that sustain life and how these pathways might be manipulated for therapeutic purposes.

## 2.4 TECHNIQUES FOR DNA AND RNA ANALYSIS

Molecular biology relies heavily on the ability to isolate and analyze DNA and RNA, the primary carriers of genetic information. DNA and RNA extraction techniques are foundational, as they allow scientists to isolate genetic material from cells for further analysis. Once extracted, the genetic material can be analyzed through methods such as Polymerase Chain Reaction (PCR) and quantitative PCR (qPCR), which amplify and quantify specific DNA or RNA sequences. PCR is used extensively in research, diagnostics, and forensics, enabling the study of genes, detection of pathogens, and identification of genetic mutations. These techniques form the backbone of genetic research and are essential tools in molecular biology labs.

## 2.5 PROTEIN PURIFICATION AND ANALYSIS

Proteins are the workhorses of the cell, carrying out a variety of functions from catalyzing reactions to signaling between cells. Purifying and analyzing proteins is fundamental for understanding their roles in biological systems. Protein purification techniques include methods such as affinity chromatography, ion-exchange chromatography, and size-exclusion chromatography, each designed to separate proteins based on their properties. Once purified, proteins are analyzed by methods such as Western blotting and enzyme-linked immunosorbent assay (ELISA) to determine their concentration, structure, and activity. These techniques are essential in drug discovery, where understanding protein function and interaction with drugs is crucial.

## 2.6 GEL ELECTROPHORESIS AND BLOTTING TECHNIQUES

Gel electrophoresis is a technique used to separate molecules based on their size and charge. In molecular biology, it is commonly used to analyze DNA, RNA, and proteins. During gel electrophoresis, samples are loaded onto a gel and subjected to an electric field, which causes molecules to migrate through the gel. Smaller molecules move faster, allowing for the separation and visualization of different fragments. After electrophoresis, blotting techniques like Southern, Northern, and Western blotting are employed to detect specific DNA, RNA, or proteins, respectively. These methods are critical for identifying genetic sequences, studying gene expression, and detecting specific proteins in complex samples.

## 2.7 MASS SPECTROMETRY IN PROTEOMICS

Mass spectrometry is a powerful tool in proteomics, the large-scale study of proteins. It allows for the identification and quantification of proteins in a sample by measuring their mass-to-charge ratio. In mass spectrometry, proteins are ionized, and their masses are detected, allowing researchers to identify proteins and analyze post-translational modifications, such as phosphorylation or glycosylation, which are important in regulating protein function.

This technique has become indispensable in studying cellular signaling pathways, disease biomarkers, and drug targets, providing a comprehensive view of the protein landscape within a cell or organism.

## 2.8 SPECTROSCOPY TECHNIQUES IN BIOCHEMISTRY

Spectroscopy techniques are essential for studying the interactions and properties of biomolecules. Ultraviolet-visible (UV-Vis) spectroscopy, fluorescence spectroscopy, and nuclear magnetic resonance (NMR) spectroscopy are commonly used in biochemistry.

UV-Vis spectroscopy measures the absorbance of light by molecules, providing information about concentration and reaction rates. Fluorescence spectroscopy is used to study the structure and dynamics of proteins and nucleic acids by observing their fluorescent emissions.

NMR spectroscopy is a powerful technique for determining the three-dimensional structure of proteins and nucleic acids. These techniques are indispensable for analyzing biomolecular interactions, studying reaction mechanisms, and characterizing molecular structures.

## 2.9 CRISPR-CAS9 AND GENOME EDITING

CRISPR-Cas9 technology has revolutionized molecular biology by enabling precise genome editing. This system utilizes a guide RNA to target a specific DNA sequence and the Cas9 enzyme to cut the DNA at that location. Researchers can then make targeted changes, such as inserting or deleting genes.

CRISPR has vast applications in genetic engineering, gene therapy, and functional genomics. It allows scientists to study gene function by creating gene knockouts and holds potential for treating genetic diseases by correcting mutations. CRISPR-Cas9 exemplifies how advancements in molecular biology techniques can have transformative impacts on both research and medicine.

## 2.10 BIOINFORMATICS AND COMPUTATIONAL BIOLOGY

As the complexity of molecular biology data increases, bioinformatics and computational biology have become essential tools for managing and analyzing large datasets. Bioinformatics involves the use of software and algorithms to analyze DNA, RNA, and protein sequences, enabling researchers to identify gene functions, predict protein structures, and understand genetic variations. Computational biology integrates mathematical models to simulate biochemical processes and molecular interactions, providing insights into cellular function and disease mechanisms. These approaches are essential for handling high-throughput data from sequencing and proteomics studies, helping to unlock new understandings in genomics, transcriptomics, and systems biology.

## 2.11 CONCLUSION

The principles and techniques of biochemistry and molecular biology have revolutionized our understanding of life at the molecular level, enabling scientists to explore the fundamental processes that govern biological function. From the structure-function relationship of biomolecules and the central dogma of molecular biology to cutting-edge techniques like CRISPR-Cas9 and bioinformatics, these methods have opened up new frontiers in scientific research. As these techniques continue to advance, they hold the potential to address some of the most pressing challenges in healthcare, environmental science, and biotechnology. Mastering these foundational principles and techniques is essential for researchers, as they provide the tools needed to uncover the complexities of life and drive innovation across diverse scientific disciplines.

# CHAPTER 3
# CELL CULTURE TECHNIQUES

Cell culture techniques are pivotal in modern biological and medical research, allowing scientists to grow and study cells in a controlled laboratory environment. By providing specific nutrients, conditions, and environments, cell culture enables researchers to investigate cellular behaviors, responses to drugs, gene expression, and interactions with pathogens. These techniques play an essential role in areas like cancer research, drug discovery, regenerative medicine, and toxicology.

## 3.1 INTRODUCTION TO CELL CULTURE

Cell culture refers to the process of growing cells outside their natural environment, typically in a petri dish, flask, or multi-well plate with a controlled nutrient medium. In this artificial environment, cells can be studied in isolation from the rest of the organism, allowing researchers to observe specific cellular processes and responses. Cell culture is a cornerstone of biomedical research, enabling advancements in understanding disease mechanisms, drug development, and tissue engineering.

## 3.2 TYPES OF CELL CULTURES

Cell cultures can be broadly classified into two main categories: primary cultures and cell lines.

### 3.2.1 Primary Culture

Primary cell culture is the *ex vivo* culture of cells freshly obtained from a multicellular organism, as opposed to the culture of immortalized cell lines. In general, primary cell cultures are considered more representative of *in vivo* tissues than cell lines, and this is recognized legally in some countries such as the UK (Human Tissue Act 2004). However, primary cells require adequate substrate and nutrient conditions to thrive and after a certain number of divisions they acquire a senescent phenotype, leading to irreversible cell cycle arrest. The generation of cell lines stems from these two reasons. Primary cells can become immortalized either spontaneously (e.g. HeLa cells) or by genetic modification (e.g. HEK cells), at which point they become cell lines which can be subcultured indefinitely.

Because of their requirements for viability, primary cell cultures did not become widespread until the 2000s. These cultures present several advantages over cell lines, including a better representation of the cellular heterogeneity of tissues, a more faithful transcriptomic and proteomic profile (especially when cultured in 3D) and more realistic functional responses, including drug responses. In contrast, immortalized cell lines are known to become homogeneous through the natural selection of specific subpopulations, to undergo genetic drift and to acquire genetic aberrations. In many cases, cell lines have been misidentified, contaminated with other cells or infected with *Mycoplasma*, small intracellular bacteria that went undetected for decades.

When whole or partial tissues are isolated and maintained *ex vivo*, the procedure is termed primary tissue culture. More specific terms include organotypic culture, tissue slices and explants.

Neuronal primary cell cultures are cells collected from the brain of an organism. For example they can be used when examining substances effect on cell viability, which can further on be potential treatments for brain deficits.

## Monolayer Cultures

"Monolayer cultures" refer to cell cultures where cells are grown in a single, flat layer on the surface of a culture dish or substrate. In a monolayer culture, cells adhere to the substrate and spread out in a two-dimensional arrangement. This type of cell culture is commonly used in laboratory settings for various purposes, including research, drug testing, and biotechnology.

Key Features of Monolayer Cultures Include:

- **Two-Dimensional Growth:** Cells in monolayer cultures grow in a single plane, adhering to the surface of the culture vessel. This flat arrangement allows for easy observation and manipulation of individual cells.
- **Adherence to Substrate:** The cells attach to the surface of the culture dish or flask, and their growth and behavior can be influenced by the characteristics of the substrate. On the other words In cell culture, adherence to substrate describes a cell's capacity to adhere to a surface and proliferate. Many elements, including surface energy, substrate topography, and roughness, mediate the process of cell attachment. The study of artificial polymer surfaces with varying chemical, topological, and mechanical cues that regulate cell activities has focused attention on the interaction between external surfaces and cells. In a study that was published in the journal RSC Advances in 2021, the impact of roughness and surface energy on cell adhesion and growth was examined. The most advantageous circumstances for effective cell adhesion, development, and proliferation were discovered by the study to be moderate surface energy and intermediate roughness ratio.

- **Cell Proliferation:** Cells in monolayer cultures can undergo cell division and proliferation. This feature is crucial for experimental studies and the production of a larger number of cells for subsequent analyses.
- **Observation and Imaging:** The two-dimensional nature of monolayer cultures makes it convenient for microscopic observation and imaging. Researchers can easily visualize the cells, study their morphology, and monitor changes over time.
- **Cell Differentiation:** Depending on the cell type and culture conditions, monolayer cultures can be used to induce cell differentiation. This is particularly important in studying developmental processes and tissue-specific functions.

### Monolayer Cultures for the Personalized Therapy in Aggressive Thyroid Cancer of Follicular Origin

Endocrine cancer with the highest incidence is thyroid cancer (TC). Differentiated thyroid cells (DTC) that originate from follicular thyroid cells account for over 90% of total thyroid cells (TC). Papillary TC (PTC), follicular TC (FTC), and Hürthle cell TC are examples of DTC. One percent of TC is anaplastic TC (ATC), which accounts for 15–40% of TC deaths.

Mortality is one of the biggest obstacles to current treatment techniques against aggressive DTC or ATC. These strategies are not entirely effective against these conditions. Recent years have seen advancements in our knowledge of the molecular and genetic underpinnings of TC development as well as the introduction of novel medications, such as tyrosine kinase inhibitors (TKIs), which target the oncogenic or signaling kinases linked to cellular proliferation.

Preclinical models have made use of thyroid cell lines that were isolated from tumoral cells and selected for their high rate of proliferation in vitro. As a result of their adaptation to in vitro growth circumstances, these cells actually lose the distinctive characteristics of the original tumor. Because of these factors, there are significant restrictions on the usage of these cell lines. More recently, monolayer cultures of human primary cells have been created, and their biological behavior has been studied. Furthermore, whereas human primary cell cultures may now be created from samples of fine-needle aspiration citology from aggressive dedifferentiated DTC or ATC, primary TC cells were previously only obtained by surgical biopsies. Without the use of useless medications, testing several TKIs in vitro on individual patients can aid in the development of novel, individualized treatments.

### Limitations of Monolayer Culture and Motivations

Scientists are investigating novel models that can more accurately mimic the structure and function of human organs because to the limitations of monolayer culture settings. Protocol improvements in recent times have led to the creation of three-dimensional (3D) organ-like architectures known as "organoids," which are able to exhibit properties of their corresponding real organs, such as morphological features, functional activities, and individual responses to particular pathogens.

### Cell Culture Protocol

For in vitro investigations to be conducted correctly, the cell culture protocol for a particular cell line must be optimized. The best culture conditions for different cell lines can differ significantly due to the heterogeneity of germ cell malignancies.

## 3.2.2 Cell Lines

Unlike primary cultures, cell lines are derived from primary cells that have undergone modifications, allowing them to proliferate indefinitely. This can occur through a process called "transformation," where cells acquire mutations that enable them to bypass normal cellular aging processes. Immortalized cell lines are easier to maintain and provide a consistent model for experiments. However, they may exhibit genetic and behavioral differences from primary cells, which can sometimes limit their applicability to certain research areas.

## 3.2.3 3D Cell Culture

A 3D cell culture is an artificially created environment in which biological cells are permitted to grow or interact with their surroundings in all three dimensions. Unlike 2D environments (e.g., a Petri dish), a 3D cell culture allows cells in vitro to grow in all directions, similar to how they would in vivo. These three-dimensional cultures are usually grown in bioreactors, small capsules in which the cells can grow into spheroids, or 3D cell colonies. Approximately 300 spheroids are usually cultured per bioreactor.

### Background

3D cell cultures have been used in research for several decades. One of the first recorded approaches for their development was at the beginning of the 20th century, with the efforts of Alexis Carrel to develop methods for prolonged in vitro tissue cultures. Early studies in the 80's, led by Mina Bissell from the Lawrence Berkeley National Laboratory, highlighted the importance of 3D techniques for creating accurate in vitro culturing models. This work focused on the importance of the extracellular matrix and the ability of cultures in artificial 3D matrices to produce physiologically relevant multicellular structures, such as acinar structures in healthy and cancerous breast tissue models. These techniques have been applied to in vitro disease models used to evaluate cellular responses to pharmaceutical compounds.

Eric Simon, in a 1988 NIH SBIR grant report, showed that electrospinning could be used to produce nano- and submicron-scale polystyrene and polycarbonate fibrous mats (now known as scaffolds) specifically intended for use as in vitro cell substrates. This early use of electrospun fibrous lattices for cell culture and tissue engineering showed that various cell types including Human Foreskin Fibroblasts (HFF), transformed Human Carcinoma (HEp-2), and Mink Lung Epithelium (MLE) would adhere to and proliferate upon the fibers. It was noted that as opposed to the flattened morphology typically seen in 2D culture, cells grown on the electrospun fibers exhibited a more histotypic rounded 3-dimensional morphology generally observed in vivo.

3D cell culture, by emulating essential aspects of the in vivo environment, including interactions between cells and the extracellular matrix, allows for the faithful recreation of structural architecture and specialized functions in normal tissues or tumors in a laboratory setting. This approach authentically models the conditions and processes of living tissues, producing responses akin to those observed in vivo. Since its inception in the 1970s, 3D cell culture has provided significant insights into the mechanisms regulating tissue homeostasis and cancer. Moreover, it has expedited translational research in the realms of cancer biology and tissue engineering.

## Properties

In living tissue, cells exist in 3D microenvironments with intricate cell-cell and cell-matrix interactions and complex transport dynamics for nutrients and cells. Standard 2D, or monolayer, cell cultures are inadequate representations of this environment, which often makes them unreliable predictors of in vivo drug efficacy and toxicity.

3D spheroids more closely resemble in vivo tissue in terms of cellular communication and the development of extracellular matrices. These matrices help the cells to be able to move within their spheroid similar to the way cells would move in living tissue. The spheroids are thus improved models for cell migration, differentiation, survival, and growth.

Furthermore, 3D cell cultures provide more accurate depiction of cell polarization, since in 2D, the cells can only be partially polarized. Moreover, cells grown in 3D exhibit different gene expression than those grown in 2D.

The third dimension of cell growth provides more contact space for mechanical inputs and for cell adhesion, which is necessary for integrin ligation, cell contraction and even intracellular signalling. Normal solute diffusion and binding to effector proteins (like growth factors and enzymes) is also reliant on the 3D cellular matrix, so it is critical for the establishment of tissue scale solute concentration gradients

For the purposes of drug toxicology screening, it is much more useful to test gene expression of in vitro cells grown in 3D than 2D, since the gene expression of the 3D spheroids will more closely resemble gene expression in vivo. Lastly, 3D cell cultures have greater stability and longer lifespans than cell cultures in 2D. This means that they are more suitable for long-term studies and for demonstrating long-term effects of the drug. 3D environments also allow the cells to grow undisturbed. In 2D, the cells must undergo regular trypsinization to provide them with sufficient nutrients for normal cell growth. 3D spheroids have been cultured in a lab setting for up to 302 days while still maintaining healthy, non-cancerous growth.

In the interdisciplinary research of biology and aerospace, the 3D printed-scaffolds are also being used for protecting cells from the effect of gravity during the launching.

## Classification of 3D Culture Methods

There are a large number of commercially available culturing tools that claim to provide the advantages of 3D cell culture. In general, the platforms can be classified in two types of 3D culturing methods: scaffold techniques and scaffold-free techniques.

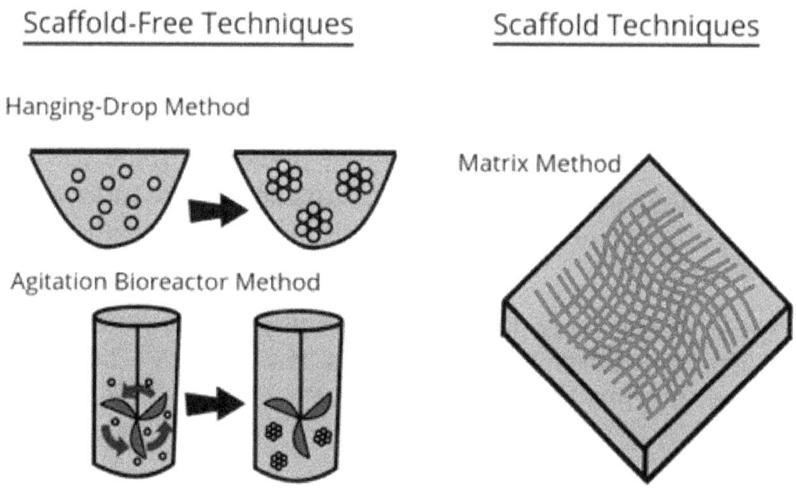


**Fig 1: A Model Showing Three Examples of Techniques Used for Culturing Cells in a 3D Environment**

## Scaffold Techniques

Scaffold techniques include the use of solid scaffolds, hydrogels and other materials. In a recent study potentiality of human CD34+ stem cells explored by generating *in vitro* agarose gel 3D model to understand the bone ossification process. Scaffolds can be used to generate microtissue 3D model by culturing fibroblasts outside of tumour cells, mimicking the tumor stroma interaction.

The effectiveness of scaffolds in various applications, particularly in tissue engineering, is significantly impacted by factors such as pore distribution, exposed surface area, and porosity. The quantity and arrangement of these elements influence both the depth and rate at which cells penetrate the scaffold volume, the structure of the resulting extracellular matrix, and ultimately, the success of the regenerative process. Scaffolds can be produced with diverse architectures depending on the manufacturing method, leading to either random or precisely designed pore distribution. Recently, advanced computer-controlled rapid prototyping techniques have been employed to create scaffolds with well-organized geometries.

### Hydrogels

As the natural extracellular matrix (ECM) is important in the survival, proliferation, differentiation and migration of the cells, different hydrogel matrices mimicking natural ECM structure are considered as potential approaches towards in vivo -like cell culturing. Hydrogels are composed of interconnected pores with high water retention, which enables efficient transport of e.g. nutrients and gases. Several different types of hydrogels from natural and synthetic materials are available for 3D cell culture, including e.g. animal ECM extract hydrogels, protein hydrogels, peptide hydrogels, polymer hydrogels, and wood-based nanocellulose hydrogel.

The approach to crafting the optimal ECM replica relies on the specific characteristics of the culture in question and typically involves employing diverse and independent chemical processes. For example, the utilization of photolabile chemistries can lead to the erosion of specific regions within a gel, and subsequently exposing these areas allows for the application of adhesive ligands, promoting cell adhesion and migration. The development of more intricate frameworks is anticipated, comprising interwoven networks of chemistries under the control of both cells and users. In essence, there is no singular network capable of faithfully emulating the intricate ECM of every tissue type. However, a thoughtful integration of bioinspired cues into synthetic gels holds the potential to yield resilient and versatile scaffolds applicable across various cell culture systems.

### Scaffold-Free Techniques

Scaffold free techniques employ another approach independent from the use scaffold. Scaffold-free methods include e.g. the use of low adhesion plates, hanging drop plates, micropatterned surfaces, and rotating bioreactors, magnetic levitation, and magnetic 3D bioprinting.

### Spheroids

Spheroids are a type of three-dimensional cell modeling that better simulate a live cell's environmental conditions compared to a two-dimensional cell model, specifically with the reactions between cells and the reactions between cells and the matrix. Spheroids are useful in the study of changing physiological characteristics of cells, the difference in the structure of healthy cells and tumor cells, and the changes cells undergo when forming a tumor.

Spheroids co-cultured with tumor and healthy cells were used to simulate how cancerous cells interact with normal cells. Spheroids can also be co-cultured with fibroblasts to mimic tumor-stroma interaction. Spheroids can be grown with a few different methods. One common method is to use low cell adhesion plates, typically a 96 well plate, to mass-produce spheroid cultures, where the aggregates form in the rounded bottom of the cell plates. Spheroids can also be cultured using the hanging drop method involving forming cell aggregates in drops that hang from the surface of a cell plate.

Other methods under investigation include the use of rotating wall vessel bioreactors, which spins and cultures the cells when they are constantly in free fall and forms aggregates in layers Recently, some protocols have been standardized to produce uniform and reliable spheroids. Researchers had also explored standardized, economical and reproducible methods for 3D cell culture. To improve reproducibility and transparency in spheroid experiments, an international consortium developed MISpheroID (Minimal Information in Spheroid Identity).

### Clusteroids

clusteroids are a type of three-dimensional cell modeling similar to spheroids but are distinguished by their creation method; grown as clusters of cells in an aqueous two-phase system of water-in-water Pickering emulsion using interfacial tension and osmotic shrinkage to pack the cells into dense clusters which are then cultured in a hydrogel into tissues or organoids.

In the absence of blood vessels, oxygen permeability is impaired during necrotic nucleus formation and this prevents the ex vivo use of 3D cell culture. There is an emulsion template that can overcome this problem. This approach allowed researchers to adjust the cell composition to attain the ideal conditions for promoting the synthesis of diverse angiogenic protein markers within the co-cultured clusteroids. HUVEC cells exhibit a reaction to the presence of Hep-G2 cells and their derivatives by generating endothelial cell sprouts in Matrigel, all without the external introduction of vascular endothelial growth factor (VEGF) or other agents that induce angiogenesis.

The replication of this cultivation technique is straightforward for generating various cell co-culture spheroids. The w/w Pickering emulsion template greatly aids in constructing 3D co-culture models, offering significant potential for applications in drug testing and tissue engineering.

## Bioreactors

The bioreactors used for 3D cell cultures are small plastic cylindrical chambers that are specifically engineered for the purpose of growing cells in three dimensions. The bioreactor uses bioactive synthetic materials such as polyethylene terephthalate membranes to surround the spheroid cells in an environment that maintains high levels of nutrients.

They are easy to open and close, so that cell spheroids can be removed for testing, yet the chamber is able to maintain 100% humidity throughout. This humidity is important to achieve maximum cell growth and function. The bioreactor chamber is part of a larger device that rotates to ensure equal cell growth in each direction across three dimensions. MC2 Biotek has developed a bioreactor to incubate proto-tissue that uses gas exchange to maintain high oxygen levels within the cell chamber. This is an improvement over previous bioreactors because the higher oxygen levels help the cell grow and undergo normal cell respiration.

Collaborative efforts between tissue engineering (TE) firms, academic institutions, and industrial partners can enhance the transformation of research-oriented bioreactors into efficient commercial manufacturing systems. Academic collaborators contribute foundational aspects, while industrial partners provide essential automation elements, ensuring compliance with regulatory standards and user-friendliness. Established consortia in Europe, such as REMEDI, AUTOBONE, and STEPS, focus on developing automated systems to streamline the engineering of autologous cell-based grafts. The aim is to meet regulatory criteria and ensure cost-effectiveness, making tissue-engineered products more clinically accessible and advancing the translational paradigm of TE from research to a competitive commercial field.

## Microfluidics

The utilization of microfluidic technology facilitates the generation of intricate micro-scale structures and the precise manipulation of parameters, thereby emulating the in vivo cellular milieu. The integration of microfluidic technology with 3D cell culture holds considerable potential for applications that seek to replicate in vivo tissue characteristics, notably exemplified by the evolving organ-on-a-chip system.

The various cell structures in the human body must be vascularized to receive the nutrients and gas exchange in order to survive. Similarly, 3D cell cultures in vitro require certain levels of fluid circulation, which can be problematic for dense, 3D cultures where cells may not all have adequate exposure to nutrients. This is particularly important in hepatocyte cultures because the liver is a highly vascularized organ. One study cultured hepatocytes and vascular cells together on a collagen gel scaffold between microfluidic channels, and compared growth of cells in static and flowing environments, and showed the need for models with tissues and a microvascular network. Another study showed that hanging-drop based spheroid co-culture device can be useful, generating two different cell spheroids on adjacent channels of microfluidic hanging drop device, and co-culturing spheroids with merging droplets, to monitor tumor-induced angiogenesis.

Microfluidic 3D cell culture, with its potential applications in biomedical research and tissue engineering, is an area of growing interest. However, its advancement is accompanied by several formidable challenges. One such challenge pertains to the difficulty in accessing cultured cells within microsystems, coupled with the intricate nature of sample extraction for subsequent assays.

Additionally, the development of methodologies and devices dedicated to in vivo-like cell metabolism and functions study, as well as drug discovery, represents a significant hurdle for microfluidic 3D cell culture devices. Another noteworthy impediment is the limited availability of microfabrication instrumentation in conventional biology laboratories. Moreover, the commercialization of mature and user-friendly microfluidic devices poses a substantial challenge, hindering their accessibility to biologists. Lastly, while biologists often seek high-throughput assay tools with optimal reproducibility, microfluidics encounters technical limitations in meeting these demands, despite the potential feasibility of parallel assays.

## High-Throughput Screening

Advanced development of 3D models for high-throughput screening in high density formats has recently been achievable due to technological achievements related to increased microplate density. These can be found in 384 and 1536-well formats that are cell repellent, cost effective, and amenable to fully automated screening platforms. Two options that afford 1536-well formats are available from either Greiner Bio-One using the m3D Magnetic 3D bioprinting and Corning Life Sciences which incorporates an ultra-low attachment surface coating, along with a microcavity geometry and gravity to create 3D models. Due to the rapid and affordable methods and technologies that have been developed for 3D screening, parallel high-throughput screening approaches to test isogenic pairs of oncogene related mutants versus wildtype have been enabled. Moreover, High-throughput screening techniques play a pivotal role in connecting the realms of pharmacology and toxicology within the framework of 3D cell culture.

## Pharmacology and Toxicology

A primary purpose of growing cells in 3D scaffolds and as 3D cell spheroids in vitro is to test pharmacokinetic and pharmacodynamic effects of drugs and nanomaterials in preclinical trials. Toxicology studies have shown 3D cell cultures to be nearly on par with in vivo studies for the purposes of testing toxicity of drug compounds. When comparing LD50 values for 6 common drugs: acetaminophen, amiodarone, diclofenac, metformin, phenformin, and valproic acid, the 3D spheroid values correlated directly with those from in vivo studies. Although 2D cell cultures have previously been used to test for toxicity along with in vivo studies, the 3D spheroids are better at testing chronic exposure toxicity because of their longer life spans. The matrix in 3D Spheroids causes cells to maintain actin filaments and is more relevant physiologically in cytoskeletal organization and cell polarity and shape of human cells. The three-dimensional arrangement allows the cultures to provide a model that more accurately resembles human tissue in vivo without using animal test subjects.

The current protocols for evaluating drug candidates and assessing toxicity heavily depend on outcomes derived from early-stage in vitro cell-based assays, with the expectation that these assays faithfully capture critical aspects of in vivo pharmacology and toxicology. Various in vitro designs have been fine-tuned for high throughput to enhance screening efficiency, allowing exhaustive libraries of potential pharmacologically relevant or potentially toxic molecules to undergo scrutiny for cell signals indicative of tissue damage or aligned with therapeutic objectives. Innovative approaches to multiplexed cell-based assay designs, involving the selection of specific cell types, signaling pathways, and reporters, have become standard practice.

Despite these advancements, a considerable percentage of new chemical and biological entities (NCEs/NBEs) encounter setbacks in late-stage human drug testing. Some receive regulatory "black box" warnings, while others are withdrawn from the market due to safety concerns post-regulatory approval. This recurrent pattern underscores the inadequacy of in vitro cell-based assays and subsequent preclinical in vivo studies in furnishing comprehensive pharmacological and toxicity data or reliable predictive capacity for comprehending the in vivo performance of drug candidates.

The absence of a dependable translational assay toolkit for pharmacology and toxicology contributes to the high cost and inefficiency of transitioning from initial in vitro cell-based screens to in vivo testing and subsequent clinical approvals.

Particular emphasis is placed on their capacity to retain essential cell and molecular interactions, as well as physiological parameters influencing cell phenotypes and responses to bioactive agents.

The distinctive advantages and challenges associated with these models are scrutinized, with a specific focus on their suitability for cell-based assays and their predictive capabilities, crucial for establishing accurate correlations with in vivo mechanisms of drug toxicity.

While assessing safety and efficacy, these models are well equipped to model a wide range of disease states. Each of these models has advantages and limitations that require model development and data interpretation. Public-private partnerships are critical to advance and stimulate research in this area.

## Criticisms

Existing 3D methods are not without limitations, including scalability, reproducibility, sensitivity, and compatibility with high-throughput screening (HTS) instruments. Cell-based HTS relies on rapid determination of cellular response to drug interaction, such as dose dependent cell viability, cell-cell/cell-matrix interaction, and/or cell migration, but the available assays are not optimized for 3D cell culturing. Another challenge faced by 3D cell culturing is the limited amount of data and publications that address mechanisms and correlations of drug interaction, cell differentiation, and cell-signalling in these 3D environments. None of the 3D methods have yet replaced 2D culturing on a large scale, including in the drug development process; although the number of 3D cell culturing publications is increasing rapidly, the current limited biochemical characterization of 3D tissue diminishes the adoption of new methods.

Drug-induced liver injury (DILI) stands as a primary cause of compound attrition in the pharmaceutical realm during the course of drug development. To preemptively assess the toxicity of compounds before embarking on laboratory animal testing, a range of in-vitro cell culture toxicity assays has been employed over the years. While two-dimensional (2D) in-vitro cell culture models are commonly utilized and have contributed significantly to our understanding, they frequently exhibit limitations in faithfully replicating the natural structures of in-vivo tissues. Although the most logical testing method involves humans, ethical constraints associated with human trials pose significant challenges. Consequently, there is a pressing need for enhanced human-relevant and predictive models to overcome these limitations.

The past decade has witnessed substantial endeavors aimed at advancing three-dimensional (3D) in-vitro cell culture models to better emulate in-vivo physiological conditions. The intrinsic advantages of 3D cell culture lie in its ability to represent cellular interactions akin to those in-vivo. When appropriately validated, 3D cell culture models can serve as a pivotal intermediary, bridging the gap between conventional 2D cell culture models and in-vivo animal models. This review endeavors to offer a comprehensive overview of the challenges associated with the sensitivity of biomarkers employed in detecting DILI during drug development. Additionally, it explores the potential of 3D cell culture models to address the existing gaps in the current paradigm, offering a promising avenue for more accurate toxicity assessments.

There are also problems using spheroids as a model for cancerous tissue. Although beneficial for 3D tissue culture, tumor spheroids have been criticized for being challenging or impossible to "manipulate gradients of soluble molecules in [3D spheroid] constructs, and to characterize cells in these complex gradients", unlike the paper-supported 3D cell culture for tissue-based bioassays explored by Ratmir et al.

Further challenges associated with complex 3D cell culture techniques include: imaging due to large scaffold sizes and incompatibility with many fluorescence microscopes, flow cytometry because it requires the dissociation of spheroids into a single-cell suspension, and the automation of liquid handling.

2D models cannot study cell-cell and cell-matrix interactions. As a result of the scarcity of preclinical models relevant to 2D cultures, 3D culture provides a pathophysiological microenvironment and has the potential to play a role in cancer drug discovery.

Tissue engineering requires 3D cellular scaffolds. As biomaterials, various natural and synthetic polymer hydrogels have been used by scientists to design 3D scaffolds. Since this barrier is a structure that mimics the natural ECM microenvironment, synthetic scaffolds may be more useful for studying specific tumorigenic steps. Finally, it is suggested that the most suitable three-dimensional models should be carefully selected according to specific targets.

## 3.3 ESSENTIAL EQUIPMENT FOR CELL CULTURE

Cell culture techniques require specialized equipment to maintain sterile conditions and provide the necessary environment for cell growth.

### 3.3.1 Laminar Flow Cabinet

A laminar flow cabinet or tissue culture hood is a partially enclosed bench work surface designed to prevent contamination of biological samples, semiconductor wafers, or any particle sensitive materials. Air is drawn through a HEPA filter and blown in a very smooth laminar flow in a narrow vertical curtain, separating the interior of the cabinet from the environment around it. The cabinet is usually made of stainless steel with no gaps or joints where spores might collect.

Despite their similar appearance, a laminar flow cabinet should not to be confused with a fume hood. A laminar flow cabinet blows unfiltered exhaust air towards the worker and is not safe for work with pathogenic agents, while a fume hood maintains negative pressure with constant exhaust to protect the user, but does not protect the work materials from contamination by the surrounding environment.

A biosafety cabinet is also easily-confused with a laminar flow cabinet, but like the fume hood is primarily designed to protect the worker rather than the biological samples. This is achieved by drawing surrounding air in and exhausting it through a HEPA filter to remove potentially hazardous microorganisms.

Laminar flow cabinets exist in both horizontal and vertical configurations, and there are many different types of cabinets with a variety of airflow patterns and acceptable uses. Cabinets may have a UV-C germicidal lamp to sterilize the interior and contents before use to prevent contamination of the experiment. Germicidal lamps are usually kept on for fifteen minutes to sterilize the interior before the cabinet is used. The light must be switched off when the cabinet is being used, to limit exposure to skin and eyes as stray ultraviolet light emissions can cause cancer and cataracts.

### 3.3.2 Incubators

#### Incubator (Culture)

An incubator is a device used to grow and maintain microbiological cultures or cell cultures. The incubator maintains optimal temperature, humidity and other conditions such as the $CO_2$ and oxygen content of the atmosphere inside. Incubators are essential for much experimental work in cell biology, microbiology and molecular biology and are used to culture both bacterial and eukaryotic cells.

An incubator is made up of a chamber with a regulated temperature. Some incubators also regulate humidity, gas composition, or ventilation within that chamber.

The simplest incubators are insulated boxes with an adjustable heater, typically going up to 60 to 65°C (140 to 149°F), though some can go slightly higher (generally to no more than 100°C). The most commonly used temperature both for bacteria such as the frequently used E. coli as well as for mammalian cells is approximately 37°C (99°F), as these organisms grow well under such conditions. For other organisms used in biological experiments, such as the budding yeast Saccharomyces cerevisiae, a growth temperature of 30°C (86°F) is optimal.

More elaborate incubators can also include the ability to lower the temperature (via refrigeration), or the ability to control humidity or $CO_2$ levels. This is important in the cultivation of mammalian cells, where the relative humidity is typically >80% to prevent evaporation and a slightly acidic pH is achieved by maintaining a $CO_2$ level of 5%.

From aiding in hatching chicken eggs to enabling scientists to understand and develop vaccines for deadly viruses, the laboratory incubator has seen numerous applications over the years it has been in use. The incubator has also provided a foundation for medical advances and experimental work in cellular and molecular biology.

While many technological advances have occurred since the primitive incubators first used in ancient Egypt and China, the main purpose of the incubator has remained unchanged: to create a stable, controlled environment conducive to research, study, and cultivation.

The earliest incubators were invented thousands of years ago in ancient Egypt and China, where they were used to keep chicken eggs warm. Use of incubators revolutionized food production, as it allowed chicks to hatch from eggs without requiring that a hen sit on them, thus freeing the hens to lay more eggs in a shorter period of time. Both early Egyptian and Chinese incubators were essentially large rooms that were heated by fires, where attendants turned the eggs at regular intervals to ensure even heat distribution.

The incubator received an update in the 16th century when Jean Baptiste Porta drew on ancient Egyptian design to create a more modern egg incubator. While he eventually had to discontinue his work due to the Spanish Inquisition, Rene-Antoine Ferchault de Reaumur took up the challenge in the middle of the 17th century. Reaumur warmed his incubator with a wood stove and monitored its temperature using the Reaumur thermometer, another of his inventions.

In the 19th century, researchers finally began to recognize that the use of incubators could contribute to medical advancements. They began to experiment to find the ideal environment for maintaining cell culture stocks. These early incubators were simply made up of bell jars that contained a single lit candle. Cultures were placed near the flame on the underside of the jar's lid, and the entire jar was placed in a dry, heated oven.

In the late 19th century, doctors realized another practical use for incubators: keeping premature or weak infants alive. The first infant incubator, used at a women's hospital in Paris, was heated by kerosene lamps. Fifty years later, Julius H. Hess, an American physician often considered to be the father of neonatology, designed an electric infant incubator that closely resembles the infant incubators in use today.

The next innovation in incubator technology came in the 1960s, when the $CO_2$ incubator was introduced to the market. Demand came when doctors realized that they could use $CO_2$ incubators to identify and study pathogens found in patients' bodily fluids. To do this, a sample was harvested and placed onto a sterile dish and into the incubator. The air in the incubator was kept at 37 degrees Celsius, the same temperature as the human body, and the incubator maintained the atmospheric carbon dioxide and nitrogen levels necessary to promote cell growth.

At this time, incubators also began to be used in genetic engineering. Scientists could create biologically essential proteins, such as insulin, with the use of incubators. Genetic modification could now take place on a molecular level, helping to improve the nutritional content and resistance to pestilence and disease of fruits and vegetables.

Incubators serve a variety of functions in a scientific lab. Incubators generally maintain a constant temperature, however additional features are often built in. Many incubators also control humidity. Shaking incubators incorporate movement to mix cultures. Gas incubators regulate the internal gas composition. Some incubators have a means of circulating the air inside of them to ensure even distribution of temperatures. Many incubators built for laboratory use have a redundant power source, to ensure that power outages do not disrupt experiments. Incubators are made in a variety of sizes, from tabletop models, to warm rooms, which serve as incubators for large numbers of samples.

## Incubator (Egg)

An incubator is a device simulating avian incubation by keeping eggs warm at a particular temperature range and in the correct humidity with a turning mechanism to hatch them.

The common names of the incubator in other terms include *breeding / hatching machines* or *hatchers*, *setters*, and *egg breeding / equipment.*

### History

The Egyptians had a method of incubating in 400 BC, using a cylindrical building or oven that had a fire at the bottom. The eggs that were incubating were placed on an inverted cone that was partially covered in ash. The eggs were placed in a woven basket that sat on top of the ashes. The building also had a roof that allowed smoke to escape, but it kept the rain out. Egyptian egg ovens are typically brick structures in a pyramidal shape, with two internal chambers.

Controlled scientific incubation required the accurate and repeatable measurement of temperature, such as the alcohol-based thermometer proposed by the French naturalist and scientist René Antoine Ferchault de Réaumur in 1730, and the temperature scale named after him.

Réaumur used the thermometer in his design for an artificial incubator, presented to the Académie des sciences in 1747, and published in 1749 as *l'Art de faire éclore et d'élever en toute saison des oiseaux domestiques de toutes espèces, soit par le moyen de la chaleur du fumier, soit par le moyen de celle du feu ordinaire*, or "The art of hatching and rearing domestic birds of all species in all seasons, either by means of the heat of manure or by means of ordinary fire".

Lyman Byce created a coal lamp incubator in 1879. The first commercial machine was made by Hearson in the year of 1881. Ira M. Petersime of Gettysburg, Ohio, USA is credited with inventing the electric egg incubator in 1922. Shortly thereafter he and his son, Ray, began manufacturing the incubators. They were granted numerous patents on design features and subsequent improvements. The innovation of the electrically powered incubator with expanded capacity (early on Petersime offered a model holding 15,000 eggs) encouraged the growth of large scale commercial hatcheries.

### Purposes

The incubator is an apparatus that is used to regulate environmental conditions such as temperature, humidity and turning for successful hatching of the fertile eggs placed in an enclosure. It is often used for growing bacterial cultures, hatching eggs artificially, or providing suitable conditions for a chemical or biological reaction. The incubator is recorded being used to hatch bird and reptile eggs. It lets the fetus inside the egg grow without the mother needing to be present to provide the warmth. Chicken eggs are recorded to hatch after about 21 days, but other species of birds can take a longer or shorter amount of time. Incubators are also used to raise birds.

An incubator should be able to set the perfect environment and condition for an egg to incubate because it regulates the factors such as temperature, humidity, and turning the eggs when necessary, so the egg is incubated properly because it plays the role of the hen in its natural state. The incubator also allows the egg to incubate while eliminating the external threats that could possibly harm the eggs.

The modern hatchery manager's goal is to produce large numbers of uniform, robust, day-old chicks. Robustness is a health criterion, originating in the embryonic life stage of the chicken - and correlating directly with the performance and resistance of individual chicks under differing farm conditions. Incubating different species of birds at the same time is possible within the same incubator.

### Incubation Methods

In industrial incubation, there are two common used methods of incubation: single-stage and multi-stage. In single-stage incubation, the incubator contains only eggs of the same embryonic age. The advantage of single-stage incubation is that climate conditions can be adjusted according to the needs of all the growing embryo. In multi-stage incubation the setter contains eggs of different embryonic ages, usually 3 to 6 age groups. Consequently, climate conditions cannot exactly be adjusted according to the needs of all the growing embryos and a compromise has to be sought to best suit the age groups presented in the setter. In a multi-stage incubation procedure, the heat produced by the older embryos is used to heat the warmth-demanding younger embryos in the same machine. Due to different physical demand in Multi-Stage and Single-stage different equipment is needed to be successful. Many industrial producers use traditional Single-Stage machines, but REAL Single-stage is available.

### Styles

Modern incubators are electrically heated, controlled by a thermostat while some are powered by solar energy due to epileptic and erratic electricity supply. They can be used in a farmhouse, such as a large chicken-raising facilities, or they can be found in a common classroom for students to observe the egg inside and when it hatches. Some industrial incubators are large enough to hold up to 124,416 eggs, and some other styles can only hold a few eggs.

The styles of incubators include:

- Setter incubator.
- Hatcher incubator.
- Combination incubator.

## Neonatal Intensive Care Unit

A neonatal intensive care unit (NICU), also known as an intensive care nursery (ICN), is an intensive care unit (ICU) specializing in the care of ill or premature newborn infants. The NICU is divided into several areas, including a critical care area for babies who require close monitoring and intervention, an intermediate care area for infants who are stable but still require specialized care, and a step down unit where babies who are ready to leave the hospital can receive additional care before being discharged. Neonatal refers to the first 28 days of life. Neonatal care, as known as specialized nurseries or intensive care, has been around since the 1960s.

The first American newborn intensive care unit, designed by Louis Gluck, was opened in October 1960 at Yale New Haven Hospital. An NICU is typically directed by one or more neonatologists and staffed by resident physicians, nurses, nurse practitioners, pharmacists, physician assistants, respiratory therapists, and dietitians. Many other ancillary disciplines and specialists are available at larger units.

### Nursing and Neonatal Populations

Healthcare institutions have varying entry-level requirements for neonatal nurses. Neonatal nurses are registered nurses (RNs), and therefore must have an Associate of Science in Nursing (ASN) or Bachelor of Science in Nursing (BSN) degree. Some countries or institutions may also require a midwifery qualification. Some institutions may accept newly graduated RNs having passed the NCLEX exam; others may require additional experience working in adult-health or medical/ surgical nursing.

Some countries offer postgraduate degrees in neonatal nursing, such as the Master of Science in Nursing (MSN) and various doctorates. A nurse practitioner may be required to hold a postgraduate degree. The National Association of Neonatal Nurses recommends two years' experience working in a NICU before taking graduate classes.

As with any registered nurse, local licensing or certifying bodies, as well as employers, may set requirements for continuing education. There are no mandated requirements to becoming an RN in an NICU, although neonatal nurses must have certification as a neonatal resuscitation provider. Some units prefer new graduates who do not have experience in other units, so they may be trained in the specialty exclusively, while others prefer nurses with more experience.

Intensive-care nurses undergo intensive didactic and clinical orientation in addition to their general nursing knowledge in order to provide highly specialized care for critical patients. Their competencies include the administration of high-risk medications, management of high-acuity patients requiring ventilator support, surgical care, resuscitation, advanced interventions such as extracorporeal membrane oxygenation or hypothermia therapy for neonatal encephalopathy procedures, as well as chronic-care management or lower acuity cares associated with premature infants such as feeding intolerance, phototherapy, or administering antibiotics. NICU RNs undergo annual skills tests and are subject to additional training to maintain contemporary practice.

## History

The problem of premature and congenitally ill infants is not a new one. As early as the 17th and 18th centuries, there were scholarly papers published that attempted to share knowledge of interventions. It was not until 1922, however, that hospitals started grouping the newborn infants into one area, now called the neonatal intensive care unit (NICU).

Before the Industrial Revolution, premature and ill infants were born and cared for at home and either lived or died without medical intervention. In the mid-nineteenth century, the infant incubator was first developed, based on the incubators used for chicken eggs. Stephane Tarnier is generally considered to be the father of the incubator (or *isolette* as it is now known), having developed it in 1880 to attempt to keep premature infants in a Paris maternity ward warm. Other methods had been used before, but this was the first closed model; in addition, he helped convince other physicians that the treatment *helped* premature infants. France became a forerunner in assisting premature infants, in part due to its concerns about a falling birth rate. After Tarnier retired, Pierre Budin, followed in his footsteps, noting the limitations of infants in incubators and the importance of breastmilk and the mother's attachment to the child. Budin is known as the father of modern perinatology, and his seminal work *The Nursling* (*Le Nourisson* in French) became the first major publication to deal with the care of the neonate. The incubator was improved in 1890 in Marseilles by Alexandre Lion, who founded in 1891 the Œuvre Maternelle des Couveuses d'Enfants in Nice and in January 1896 in Paris.

Another factor that contributed to the development of modern neonatology was Martin Couney and his permanent installment of premature babies in incubators at Coney Island. A more controversial figure, he studied under Budin and brought attention to premature babies and their plight through his display of infants as sideshow attractions at Coney Island and the World's Fair in New York and Chicago in 1933 and 1939, respectively. Infants had also previously been displayed in incubators at the 1897, 1898, 1901, and 1904 World Fairs.

Doctors took an increasing role in childbirth from the eighteenth century onward. However, the care of newborn babies, sick or well, remained largely in the hands of mothers and midwives. Some baby incubators, similar to those used for hatching chicks, were devised in the late nineteenth century. In the United States, these were shown at commercial exhibitions, complete with babies inside, until 1931. A. Robert Bauer, at Henry Ford Hospital in Detroit, Michigan, successfully combined oxygen, heat, humidity, ease of accessibility, and ease of nursing care in 1931. It was not until after the Second World War that *special-care baby units* (*SCBUs*, pronounced 'scaboos') were established in many hospitals. In Britain, early SCBUs opened in Birmingham and Bristol, the latter set up with only £100. At Southmead Hospital, Bristol, initial opposition from obstetricians lessened after quadruplets born there in 1948 were successfully cared for in the new unit.

Incubators were expensive, so the whole room was often kept warm instead. Cross-infection between babies was greatly feared. Strict nursing routines involved staff wearing gowns and masks, constant hand-washing and minimal handling of babies. Parents were sometimes allowed to watch through the windows of the unit. Much was learned about feeding—frequent, tiny feeds seemed best—and breathing. Oxygen was given freely until the end of the 1950s, when it was shown that the high concentrations reached inside incubators caused some babies to go blind. Monitoring conditions in the incubator, and the baby itself, was to become a major area of research.

The 1960s were a time of rapid medical advances, particularly in respiratory support, that were at last making the survival of premature newborn babies a reality. Very few babies born before thirty two weeks survived and those who did often had neurological impairments. Herbert Barrie in London pioneered advances in resuscitation of the newborn. Barrie published his seminal paper on the subject in *The Lancet* in 1963. One of the concerns at this time was the worry that using high pressures of oxygen could be damaging to newborn lungs. Barrie developed an underwater safety valve in the oxygen circuit. The tubes were originally made of rubber, but these had the potential to cause irritation to sensitive newborn tracheas: Barrie switched to plastic. This new endotracheal tube, based on Barrie's design, was known as the "St Thomas's tube".

Most early units had little equipment, providing only oxygen and warmth, and relied on careful nursing and observation. In later years, further research allowed technology to play a larger role in the decline of infant mortality. The development of pulmonary surfactant, which facilitates the oxygenation and ventilation of underdeveloped lungs, has been the most important development in neonatology to date.

By the 1970s, NICUs were an established part of hospitals in the developed world. In Britain, some early units ran community programmes, sending experienced nurses to help care for premature babies at home. But increasingly technological monitoring and therapy meant special care for babies became hospital-based. By the 1980s, over 90% of births took place in hospitals. The emergency dash from home to the NICU with baby in a transport incubator had become a thing of the past, though transport incubators were still needed. Specialist equipment and expertise were not available at every hospital, and strong arguments were made for large, centralised NICUs. On the downside was the long travelling time for frail babies and for parents. A 1979 study showed that 20% of babies in NICUs for up to a week were never visited by either parent. Centralised or not, by the 1980s few questioned the role of NICUs in saving babies. Around 80% of babies born weighing less than 1.5 kg now survived, compared to around 40% in the 1960s. From 1982, pediatricians in Britain could train and qualify in the sub-specialty of neonatal medicine.

Not only careful nursing but also new techniques and instruments now played a major role. As in adult intensive-care units, the use of monitoring and life-support systems became routine. These needed special modification for small babies, whose bodies were tiny and often immature. Adult ventilators, for example, could damage babies' lungs and gentler techniques with smaller pressure changes were devised. The many tubes and sensors used for monitoring the baby's condition, blood sampling and artificial feeding made some babies scarcely visible beneath the technology. Furthermore, by 1975, over 18% of newborn babies in Britain were being admitted to NICUs. Some hospitals admitted all babies delivered by Caesarian section or under 2500 g in weight. The fact that these babies missed early close contact with their mothers was a growing concern. The 1980s saw questions being raised about the human and economic costs of too much technology, and admission policies gradually became more conservative.

NICUs now concentrate on treating very small, premature, or congenitally ill babies. Some of these babies are from higher-order multiple births, but most are still single babies born too early. Premature labour, and how to prevent it, remains a perplexing problem for doctors. Even though medical advancements allow doctors to save low-birth-weight babies, it is almost invariably better to delay such births.

Over the last 10 years or so, SCBUs have become much more "parent-friendly", encouraging maximum involvement with the babies. Routine gowns and masks are gone and parents are encouraged to help with care as much as possible. Cuddling and skin-to-skin contact, also known as "kangaroo care", are seen as beneficial for all but the frailest (very tiny babies are exhausted by the stimulus of being handled; or larger critically ill infants). Less stressful ways of delivering high-technology medicine to tiny patients have been devised: sensors to measure blood oxygen levels through the skin, for example; and ways of reducing the amount of blood taken for tests.

Some major problems of the NICU have almost disappeared. Exchange transfusions, in which all the blood is removed and replaced, are rare now. Rhesus incompatibility (a difference in blood groups) between mother and baby is largely preventable, and was the most common cause for exchange transfusion in the past. However, breathing difficulties, intraventricular hemorrhage, necrotizing enterocolitis and infections still claim many infant lives and are the focus of many new and current research projects.

The long-term outlook for premature babies saved by NICUs has always been a concern. From the early years, it was reported that a higher proportion than normal grew up with disabilities, including cerebral palsy and learning difficulties. Now that treatments are available for many of the problems faced by tiny or immature babies in the first weeks of life, long-term follow-up, and minimising long-term disability, are major research areas.

Besides prematurity and extreme low birth-weight, common diseases cared for in a NICU include perinatal asphyxia, major birth defects, sepsis, neonatal jaundice, and infant respiratory distress syndrome due to immaturity of the lungs. In general, the leading cause of death in NICUs is necrotizing enterocolitis. Complications of extreme prematurity may include intracranial hemorrhage, chronic bronchopulmonary dysplasia or retinopathy of prematurity. An infant may spend a day of observation in a NICU or may spend many months there.

Neonatology and NICUs have greatly increased the survival of very low birth-weight and extremely premature infants. In the era before NICUs, infants of birth weight less than 1,400 grams (3.1 pounds), usually about 30 weeks gestation, rarely survived. Today, infants of 500 grams (1.1 pounds) at 26 weeks have a fair chance of survival. As of 2022, the world record for the lowest gestational age newborn to survive is held by Curtis Zy-Keith Means, who was born on July 5, 2020 in the United States, at 21 weeks and 1 day gestational age, weighing 420 grams (0.93 pounds).

The NICU environment provides challenges as well as benefits. Stressors for the infants can include continual light, a high level of noise, separation from their mothers, reduced physical contact, painful procedures, and interference with the opportunity to breastfeed. There have been very few studies investigating noise reduction interventions in the NICU and it remains uncertain what their effects could be on babies' growth and development. An NICU can be stressful for the staff as well. A special aspect of NICU stress for both parents and staff is that infants may survive, but with damage to the brain, lungs, or eyes. When parents arrive at the NICU, they will have the availability to tour the unit and orientation to the various areas and equipment. This tour includes information on the different types of equipment used in the NICU, such as incubators, monitors, and ventilators, and how they help to support the health and well-being of the babies. Parental orientation to the NICU is essential in reducing parental anxiety and improving satisfaction with care.

Effective communication is critical in the NICU. Parents will be given information on who their primary point of contact is and how they can communicate with the medical staff caring for their baby. Parents should ask questions when given tour of the NICU just in case anything was misunderstood. The gynecologic and neonatal nursing found that effective communication between health care providers and parents in the NICU is critical for promoting parental involvement and reducing stress NICU rotations are essential aspects of pediatric and obstetric residency programs, but NICU experience is encouraged by other specialty residencies, such as family practice, surgery, pharmacy, and emergency medicine.

## Equipment

An incubator (or *isolette* or *humidicrib*) is an apparatus used to maintain environmental conditions suitable for a neonate (newborn baby). It is used in preterm births or for some ill full-term babies.

Additional items of equipment used to evaluate and treat sick neonates include:

- **Blood Pressure Monitor:** The blood pressure monitor is a machine that's connected to a small cuff which is wrapped around the arm or leg of the patient. This cuff automatically takes the blood pressure and displays the data for review by care providers. Oxygen hood: This is a clear box that fits over the baby's head and supplies oxygen. This is used for babies who can still breathe but need some respiratory support.
- **Ventilator:** This is a breathing machine that delivers air to the lungs. Babies who are severely ill will receive this intervention. Typically, the

ventilator takes the role of the lungs while treatment is administered to improve lung and circulatory function.

Possible functions of a neonatal incubator are:

- **Oxygenation**, through oxygen supplementation by head hood or nasal cannula, or even continuous positive airway pressure (CPAP) or mechanical ventilation. Infant respiratory distress syndrome is the leading cause of death in preterm infants, and the main treatments are CPAP, in addition to administering pulmonary surfactant and stabilizing the blood sugar, blood salts, and blood pressure.
- **Observation:** Modern neonatal intensive care involves sophisticated measurement of temperature, respiration, cardiac function, oxygenation, and brain activity.
- **Protection from Cold Temperature, Infection, Noise, Drafts and Excess Handling:** Incubators may be described as bassinets enclosed in plastic, with climate control equipment designed to keep them warm and limit their exposure to germs.
- **Maintaining** fluid balance by providing fluid and keeping a high air humidity to prevent too great a loss from skin and respiratory evaporation.

A *transport incubator* is an incubator in a transportable form, and is used when a sick or premature baby is moved, e.g., from one hospital to another, as from a community hospital to a larger medical facility with a proper neonatal intensive-care unit. It usually has a miniature ventilator, cardio-respiratory monitor, IV pump, pulse oximeter, and oxygen supply built into its frame.

### Pain Management

Many parents with newborns in the NICU have expressed that they would like to learn more about what types of pain their infants are feeling and how they can help relieve it. Another main worry that was mentioned was the long-term effects of their pain, and whether it mentally affect the child in the future, or even affect the relationship they have with their parents. There are multiple ways to manage pain for infants. If the mother is able to help, holding the infant in kangaroo position or breastfeeding can help calm the baby before a procedure is done. Other simple things that can help ease pain include: allowing the infant to suck on a gloved finger, gently binding the limbs in a flexed position, and creating a quiet and comfortable environment.

### Levels of Care

The concept of designations for hospital facilities that care for newborn infants according to the level of complexity of care provided was first proposed in the United States in 1976. Levels in the United States are designated by the guidelines published by the American Academy of Pediatrics In Britain, the guidelines are issued by The British Association of Perinatal Medicine (BAPM), and in Canada, they are maintained by The Canadian Paediatric Society.

Neonatal care is split into categories or "levels of care". these levels apply to the type of care needed and is determined by the governing body of the area.

The definition of a neonatal intensive-care unit (NICU) according to the National Center for Statistics is a "hospital facility or unit staffed and equipped to provide continuous mechanical ventilatory support for a newborn infant". In 2012, the American Academy of Pediatric updated their policy statement delineating the different levels of neonatal care. One major difference in the 2012 updated policy statement from the AAP compared to the 2004 policy statement is the removal of subspeciality nurseries for levels II and III with the addition of a level IV NICU. The four distinct levels of neonatal care defined in the most recent policy statement from the AAP are:

- **Level I**, Well newborn nursery.
- **Level II**, Special care nursery.
- **Level III**, Neonatal intensive-care unit (NICU).
- **Level IV**, Regional neonatal intensive-care unit (Regional NICU).

Level I units are typically referred to as the well baby nursery. Well newborn nurseries have the capability to provide neonatal resuscitation at every delivery; evaluate and provide postnatal care to healthy newborn infants; stabilize and provide care for infants born at 35 to 37 weeks' gestation who remain physiologically stable; and stabilize newborn infants who are ill and those born less than 35 weeks' gestation until transfer to a facility that can provide the appropriate level of neonatal care. Required provider types for well newborn nurseries include pediatricians, family physicians, nurse practitioners, and other advanced practice registered nurses.

Previously, Level II units were subdivided into 2 categories (level IIA & level IIB) on the basis of their ability to provide assisted ventilation including continuous positive airway pressure. Level II units are also known as special care nurseries and have all of the capabilities of a level I nursery. In addition to providing level I neonatal care, Level II units are able to:

- Provide care for infants born ≥32-week gestation and weighing ≥1500 g who have physiologic immaturity or who are moderately ill with problems that are expected to resolve rapidly and are not anticipated to need subspecialty services on an urgent basis.

- Provide care for infants who are feeding and growing stronger or convalescing after intensive care.
- Provide mechanical ventilation for a brief duration (<24 h) or continuous positive airway pressure.
- Stabilize infants born before 32-week gestation and weighing less than 1500 g until transfer to a neonatal intensive-care facility.
- Level II nurseries are required to be managed and staffed by a pediatrician, however many Level II special care nurseries are staffed by neonatologists and neonatal nurse practitioners.

The 2004 AAP guidelines subdivided Level III units into 3 categories (level IIIA, IIIB & IIIC). Level III units are required to have pediatric surgeons in addition to care providers required for level II (pediatric hospitalists, neonatologists, and neonatal nurse practitioners) and level I (pediatricians, family physicians, nurse practitioners, and other advanced practice registered nurses).

Also, required provider types that must either be on site or at a closely related institution by prearranged consultative agreement include pediatric medical subspecialists, pediatric anesthesiologists, and pediatric ophthalmologists.

In addition to providing the care and having the capabilities of level I and level II nurseries, level III neonatal intensive-care units are able to:

- Provide sustained life support.
- Provide comprehensive care for infants born <32 wks gestation and weighing <1500 g.
- Provide comprehensive care for infants born at all gestational ages and birth weights with critical illness.
- Provide prompt and readily available access to a full range of pediatric medical subspecialists, pediatric surgical specialists, pediatric anesthesiologists, and pediatric ophthalmologists.
- Provide a full range of respiratory support that may include conventional and/or high-frequency ventilation and inhaled nitric oxide.
- Perform advanced imaging, with interpretation on an urgent basis, including computed tomography, MRI, and echocardiography.

The highest level of neonatal care provided occurs at regional NICUs, or Level IV neonatal intensive-care units. Level IV units are required to have pediatric surgical subspecialists in addition to the care providers required for Level III units. Regional NICUs have all of the capabilities of Level I, II, and III units. In addition to providing the highest level of care, level IV NICUs:

- Are located within an institution with the capability to provide surgical repair of complex congenital or acquired conditions.
- Maintain a full range of pediatric medical subspecialists, pediatric surgical subspecialists, and pediatric anesthesiologists at the site.
- Facilitate transport and provide outreach education.

The terminology used in the United Kingdom can be confusing because different criteria are used to designate 'special' and 'intensive' neonatal care locally and nationally. Also known as 'Special Care Baby Units' (SCBU). These look after babies who need more care than healthy newborns but are relatively stable and mature. SCBU might provide tube-feeding, oxygen therapy, antibiotics to treat infection and phototherapy for jaundice. In a SCBU, a nurse can be assigned up to four babies to care for.

Also known as 'Local Neonatal Units', these can look after babies who need more advanced support such as parenteral nutrition and continuous positive airway pressure (CPAP). Confusingly, they may also look after babies who need short-term intensive care such as mechanical ventilation. Babies who will need longer-term or more elaborate intensive care, for example extremely preterm infants, are usually transferred to a Level 3 unit. Babies in a Level 2 unit may be classified for nursing purposes as 'Special Care', 'High Dependency' (HDU) (in which a nurse will be assigned up to two babies) or 'Intensive care' (where nursing is one-to-one, or sometimes even two-to-one).

Also known as 'Neonatal Intensive Care Units' (NICU) - although Level 2 units may also have their own NICU. These look after the smallest, most premature and most unwell babies and often serve a large geographical region. Therapies such as prolonged mechanical ventilation, therapeutic hypothermia, neonatal surgery and inhaled nitric oxide are usually provided in Level 3 Units, although not every unit has access to all therapies. Some babies being cared for in Level 3 units will require less intensive treatment and will be looked after in HDU or SCBU nurseries on the same site. NHS England recommended in December 2019 that these units should care for at least 100 babies weighing less than 1.5 kg, and usually perform more than 2,000 intensive care days per year.

India has 3-tier system based on weight and gestational age of neonate. Neonates weighing more than 1800 grams or having gestational maturity of 34 weeks or more are categorized under level I care. The care consists of basic care at birth, provision of warmth, maintaining asepsis and promotion of breastfeeding. This type of care can be given at home, subcenter and primary health centre.

Neonates weighing 1200-1800 grams or having gestational maturity of 30–34 weeks are categorized under level II care and are looked after by trained nurses and pediatricians. The equipment and facilities used for this level of care include equipment for resuscitation, maintenance of thermoneutral environment, intravenous infusion, gavage feeding, phototherapy and exchange blood transfusion. This type of care can be given at first referral units, district hospitals, teaching institutions and nursing homes.

Neonates weighing less than 1200 grams or having gestational maturity of less than 30 weeks are categorized under level III care. The care is provided at apex institutions and regional perinatal centers equipped with centralized oxygen and suction facilities, servo-controlled incubators, vital signs monitors, transcutaneous monitors, ventilators, infusion pumps etc. This type of care is provided by skilled nurses and neonatologists.

### 3.3.3 Culture Vessels

Culture vessels like petri dishes, flasks, and multi-well plates are designed to hold cells and growth medium. These containers are often treated to facilitate cell attachment and promote optimal cell growth.

### 3.3.4 Microscopes

A microscope is a laboratory instrument used to examine objects that are too small to be seen by the naked eye. Microscopy is the science of investigating small objects and structures using a microscope. Microscopic means being invisible to the eye unless aided by a microscope.

There are many types of microscopes, and they may be grouped in different ways. One way is to describe the method an instrument uses to interact with a sample and produce images, either by sending a beam of light or electrons through a sample in its optical path, by detecting photon emissions from a sample, or by scanning across and a short distance from the surface of a sample using a probe. The most common microscope (and the first to be invented) is the optical microscope, which uses lenses to refract visible light that passed through a thinly sectioned sample to produce an observable image. Other major types of microscopes are the fluorescence microscope, electron microscope (both the transmission electron microscope and the scanning electron microscope) and various types of scanning probe microscopes.

#### History

Although objects resembling lenses date back 4,000 years and there are Greek accounts of the optical properties of water-filled spheres (5th century BC) followed by many centuries of writings on optics, the earliest known use of simple microscopes (magnifying glasses) dates back to the widespread use of lenses in eyeglasses in the 13th century. The earliest known examples of compound microscopes, which combine an objective lens near the specimen with an eyepiece to view a real image, appeared in Europe around 1620. The inventor is unknown, even though many claims have been made over the years. Several revolve around the spectacle-making centers in the Netherlands, including claims it was invented in 1590 by Zacharias Janssen (claim made by his son) or Zacharias' father, Hans Martens, or both, claims it was invented by their neighbor and rival spectacle maker, Hans Lippershey (who applied for the first telescope patent in 1608), and claims it was invented by expatriate Cornelis Drebbel, who was noted to have a version in London in 1619. Galileo Galilei (also sometimes cited as compound microscope inventor) seems to have found after 1610 that he could close focus his telescope to view small objects and, after seeing a compound microscope built by Drebbel exhibited in Rome in 1624, built his own improved version.

Giovanni Faber coined the name *microscope* for the compound microscope Galileo submitted to the Accademia dei Lincei in 1625 (Galileo had called it the *occhiolino* 'little eye'). René Descartes (*Dioptrique*, 1637) describes microscopes wherein a concave mirror, with its concavity towards the object, is used, in conjunction with a lens, for illuminating the object, which is mounted on a point fixing it at the focus of the mirror.

The first detailed account of the microscopic anatomy of organic tissue based on the use of a microscope did not appear until 1644, in Giambattista Odierna's *L'occhio della mosca*, or *The Fly's Eye*.

The microscope was still largely a novelty until the 1660s and 1670s when naturalists in Italy, the Netherlands and England began using them to study biology. Italian scientist Marcello Malpighi, called the father of histology by some historians of biology, began his analysis of biological structures with the lungs. The publication in 1665 of Robert Hooke's *Micrographia* had a huge impact, largely because of its impressive illustrations. Hooke created tiny lenses of small glass globules made by fusing the ends of threads of spun glass.

A significant contribution came from Antonie van Leeuwenhoek who achieved up to 300 times magnification using a simple single lens microscope. He sandwiched a very small glass ball lens between the holes in two metal plates riveted together, and with an adjustable-by-screws needle attached to mount the specimen. Then, Van Leeuwenhoek re-discovered red blood cells (after Jan Swammerdam) and spermatozoa, and helped popularise the use of microscopes to view biological ultrastructure. On 9 October 1676, van Leeuwenhoek reported the discovery of micro-organisms.

The performance of a compound light microscope depends on the quality and correct use of the condensor lens system to focus light on the specimen and the objective lens to capture the light from the specimen and form an image. Early instruments were limited until this principle was fully appreciated and developed from the late 19th to very early 20th century, and until electric lamps were available as light sources. In 1893 August Köhler developed a key principle of sample illumination, Köhler illumination, which is central to achieving the theoretical limits of resolution for the light microscope.

This method of sample illumination produces even lighting and overcomes the limited contrast and resolution imposed by early techniques of sample illumination. Further developments in sample illumination came from the discovery of phase contrast by Frits Zernike in 1953, and differential interference contrast illumination by Georges Nomarski in 1955; both of which allow imaging of unstained, transparent samples.

In the early 20th century a significant alternative to the light microscope was developed, an instrument that uses a beam of electrons rather than light to generate an image. The German physicist, Ernst Ruska, working with electrical engineer Max Knoll, developed the first prototype electron microscope in 1931, a transmission electron microscope (TEM). The transmission electron microscope works on similar principles to an optical microscope but uses electrons in the place of light and electromagnets in the place of glass lenses. Use of electrons, instead of light, allows for much higher resolution.

Development of the transmission electron microscope was quickly followed in 1935 by the development of the scanning electron microscope by Max Knoll. Although TEMs were being used for research before WWII, and became popular afterwards, the SEM was not commercially available until 1965.

Transmission electron microscopes became popular following the Second World War. Ernst Ruska, working at Siemens, developed the first commercial transmission electron microscope and, in the 1950s, major scientific conferences on electron microscopy started being held. In 1965, the first commercial scanning electron microscope was developed by Professor Sir Charles Oatley and his postgraduate student Gary Stewart, and marketed by the Cambridge Instrument Company as the "Stereoscan".

One of the latest discoveries made about using an electron microscope is the ability to identify a virus. Since this microscope produces a visible, clear image of small organelles, in an electron microscope there is no need for reagents to see the virus or harmful cells, resulting in a more efficient way to detect pathogens.

New types of scanning probe microscope have continued to be developed as the ability to machine ultra-fine probes and tips has advanced.

The most recent developments in light microscope largely centre on the rise of fluorescence microscopy in biology. During the last decades of the 20th century, particularly in the post-genomic era, many techniques for fluorescent staining of cellular structures were developed. The main groups of techniques involve targeted chemical staining of particular cell structures, for example, the chemical compound DAPI to label DNA, use of antibodies conjugated to fluorescent reporters, and fluorescent proteins, such as green fluorescent protein. These techniques use these different fluorophores for analysis of cell structure at a molecular level in both live and fixed samples.

The rise of fluorescence microscopy drove the development of a major modern microscope design, the confocal microscope. The principle was patented in 1957 by Marvin Minsky, although laser technology limited practical application of the technique. It was not until 1978 when Thomas and Christoph Cremer developed the first practical confocal laser scanning microscope and the technique rapidly gained popularity through the 1980s.

Much current research (in the early 21st century) on optical microscope techniques is focused on development of superresolution analysis of fluorescently labelled samples. Structured illumination can improve resolution by around two to four times and techniques like stimulated emission depletion (STED) microscopy are approaching the resolution of electron microscopes. This occurs because the diffraction limit is occurred from light or excitation, which makes the resolution must be doubled to become super saturated. Stefan Hell was awarded the 2014 Nobel Prize in Chemistry for the development of the STED technique, along with Eric Betzig and William Moerner who adapted fluorescence microscopy for single-molecule visualization.

X-ray microscopes are instruments that use electromagnetic radiation usually in the soft X-ray band to image objects. Technological advances in X-ray lens optics in the early 1970s made the instrument a viable imaging choice. They are often used in tomography to produce three dimensional images of objects, including biological materials that have not been chemically fixed. Currently research is being done to improve optics for hard X-rays which have greater penetrating power.

## Types

Microscopes can be separated into several different classes. One grouping is based on what interacts with the sample to generate the image, i.e., light or photons (optical microscopes), electrons (electron microscopes) or a probe (scanning probe microscopes).

Alternatively, microscopes can be classified based on whether they analyze the sample via a scanning point (confocal optical microscopes, scanning electron microscopes and scanning probe microscopes) or analyze the sample all at once (wide field optical microscopes and transmission electron microscopes).

Wide field optical microscopes and transmission electron microscopes both use the theory of lenses (optics for light microscopes and electromagnet lenses for electron microscopes) in order to magnify the image generated by the passage of a wave transmitted through the sample, or reflected by the sample. The waves used are electromagnetic (in optical microscopes) or electron beams (in electron microscopes). Resolution in these microscopes is limited by the wavelength of the radiation used to image the sample, where shorter wavelengths allow for a higher resolution.

Scanning optical and electron microscopes, like the confocal microscope and scanning electron microscope, use lenses to focus a spot of light or electrons onto the sample then analyze the signals generated by the beam interacting with the sample. The point is then scanned over the sample to analyze a rectangular region. Magnification of the image is achieved by displaying the data from scanning a physically small sample area on a relatively large screen. These microscopes have the same resolution limit as wide field optical, probe, and electron microscopes.

Scanning probe microscopes also analyze a single point in the sample and then scan the probe over a rectangular sample region to build up an image. The most common type of microscope (and the first invented) is the optical microscope. This is an optical instrument containing one or more lenses producing an enlarged

image of a sample placed in the focal plane. Optical microscopes have refractive glass (occasionally plastic or quartz), to focus light on the eye or on to another light detector. Mirror-based optical microscopes operate in the same manner. Typical magnification of a light microscope, assuming visible range light, is up to 1,250× with a theoretical resolution limit of around 0.250 micrometres or 250 nanometres. This limits practical magnification to ~1,500×. Specialized techniques (e.g., scanning confocal microscopy, Vertico SMI) may exceed this magnification but the resolution is diffraction limited.

The use of shorter wavelengths of light, such as ultraviolet, is one way to improve the spatial resolution of the optical microscope, as are devices such as the near-field scanning optical microscope.

Sarfus is a recent optical technique that increases the sensitivity of a standard optical microscope to a point where it is possible to directly visualize nanometric films (down to 0.3 nanometre) and isolated nano-objects (down to 2 nm-diameter). The technique is based on the use of non-reflecting substrates for cross-polarized reflected light microscopy.

Ultraviolet light enables the resolution of microscopic features as well as the imaging of samples that are transparent to the eye. Near infrared light can be used to visualize circuitry embedded in bonded silicon devices, since silicon is transparent in this region of wavelengths.

In fluorescence microscopy many wavelengths of light ranging from the ultraviolet to the visible can be used to cause samples to fluoresce, which allows viewing by eye or with specifically sensitive cameras.

Phase-contrast microscopy is an optical microscopic illumination technique in which small phase shifts in the light passing through a transparent specimen are converted into amplitude or contrast changes in the image. The use of phase contrast does not require staining to view the slide. This microscope technique made it possible to study the cell cycle in live cells.

The traditional optical microscope has more recently evolved into the digital microscope. In addition to, or instead of, directly viewing the object through the eyepieces, a type of sensor similar to those used in a digital camera is used to obtain an image, which is then displayed on a computer monitor. These sensors may use CMOS or charge-coupled device (CCD) technology, depending on the application.

Digital microscopy with very low light levels to avoid damage to vulnerable biological samples is available using sensitive photon-counting digital cameras. It has been demonstrated that a light source providing pairs of entangled photons may minimize the risk of damage to the most light-sensitive samples. In this application of ghost imaging to photon-sparse microscopy, the sample is illuminated with infrared photons, each of which is spatially correlated with an entangled partner in the visible band for efficient imaging by a photon-counting camera.

The two major types of electron microscopes are transmission electron microscopes (TEMs) and scanning electron microscopes (SEMs). They both have series of electromagnetic and electrostatic lenses to focus a high energy beam of electrons on a sample. In a TEM the electrons pass through the sample, analogous to basic optical microscopy. This requires careful sample preparation, since electrons are scattered strongly by most materials. The samples must also be very thin in order for the electrons to pass through it. Cross-sections of cells stained with osmium and heavy metals reveal clear organelle membranes and proteins such as ribosomes. With a 0.1nm level of resolution, detailed views of viruses (20–300nm) and a strand of DNA (2nm in width) can be obtained. In contrast, the SEM has raster coils to scan the surface of bulk objects with a fine electron beam. Therefore, the specimen do not necessarily need to be sectioned, but coating with a nanometric metal or carbon layer may be needed for nonconductive samples. SEM allows fast surface imaging of samples, possibly in thin water vapor to prevent drying.

The different types of scanning probe microscopes arise from the many different types of interactions that occur when a small probe is scanned over and interacts with a specimen. These interactions or modes can be recorded or mapped as function of location on the surface to form a characterization map. The three most common types of scanning probe microscopes are atomic force microscopes

(AFM), near-field scanning optical microscopes (NSOM or SNOM, scanning near-field optical microscopy), and scanning tunneling microscopes (STM). An atomic force microscope has a fine probe, usually of silicon or silicon nitride, attached to a cantilever; the probe is scanned over the surface of the sample, and the forces that cause an interaction between the probe and the surface of the sample are measured and mapped. A near-field scanning optical microscope is similar to an AFM but its probe consists of a light source in an optical fiber covered with a tip that has usually an aperture for the light to pass through. The microscope can capture either transmitted or reflected light to measure very localized optical properties of the surface, commonly of a biological specimen. Scanning tunneling microscopes have a metal tip with a single apical atom; the tip is attached to a tube through which a current flows. The tip is scanned over the surface of a conductive sample until a tunneling current flows; the current is kept constant by computer movement of the tip and an image is formed by the recorded movements of the tip.

Scanning acoustic microscopes use sound waves to measure variations in acoustic impedance. Similar to Sonar in principle, they are used for such jobs as detecting defects in the subsurfaces of materials including those found in integrated circuits. On February 4, 2013, Australian engineers built a "quantum microscope" which provides unparalleled precision.

Mobile app microscopes can optionally be used as optical microscope when the flashlight is activated. However, mobile app microscopes are harder to use due to visual noise, are often limited to 40x, and the resolution limits of the camera lens itself.

## 3.4 CELL CULTURE MEDIUM

The culture medium provides essential nutrients, growth factors, hormones, and salts necessary for cell survival and proliferation. Most media are supplemented with serum, such as fetal bovine serum (FBS), which supplies additional growth factors and nutrients. Different cell types require specific media formulations that cater to their unique requirements. Careful selection of media composition is essential for successful cell growth, as changes in nutrients and growth factors can alter cell behavior.

## 3.5 STERILE TECHNIQUES IN CELL CULTURE

Maintaining sterility is a crucial aspect of cell culture, as contamination with bacteria, fungi, or viruses can rapidly spoil cultures and compromise experimental results. Key sterile practices include working within a laminar flow hood, using sterilized equipment, wearing protective clothing, and avoiding unnecessary movements that could disturb the sterile air within the hood. In addition, regular testing for contamination is often conducted to ensure that cultures remain clean.

## 3.6 CRYOPRESERVATION OF CELLS

Cryopreservation is the process of freezing cells to preserve them for long periods. This technique is essential for storing cell lines, as it allows researchers to maintain a backup of their cultures without continuous sub-culturing. In cryopreservation, cells are treated with a cryoprotectant (such as dimethyl sulfoxide, or DMSO) to prevent ice formation, then frozen at ultra-low temperatures (usually in liquid nitrogen). When needed, cells can be thawed and re-cultured, enabling researchers to preserve cell viability and maintain consistency in experiments.

## 3.7 PASSAGING AND SUBCULTURING CELLS

To prevent cells from becoming overcrowded in their culture vessel, researchers must perform passaging or subculturing. This involves transferring a portion of cells to a new vessel to allow continued growth. Passaging helps maintain cell health, as overconfluent cultures can lead to changes in cell behavior and morphology. Cells are typically detached using trypsin, an enzyme that breaks down proteins that keep the cells adhered to the vessel surface, before being transferred to fresh media. Maintaining regular passage schedules is critical in ensuring cell viability and consistency.

## 3.8 APPLICATIONS OF CELL CULTURE TECHNIQUES

Cell culture techniques have a wide range of applications in scientific research and industrial processes.

- **Drug Development and Testing:** Cell culture allows scientists to test the efficacy and toxicity of new drugs on specific cell types before moving to animal models or human trials. High-throughput screening of potential drug compounds is commonly performed on cell cultures, providing a rapid and cost-effective means of identifying promising candidates.
- **Cancer Research:** Cancer cells are often cultured to study their biology and response to treatments. Cell culture techniques enable researchers to investigate the molecular mechanisms of cancer development, identify genetic mutations, and test the effects of potential cancer therapies.
- **Gene Expression Studies:** Cell cultures are frequently used to study gene expression and regulation. By manipulating the genetic material within cultured cells, scientists can analyze how genes function and regulate cellular processes, leading to insights into diseases and potential genetic therapies.
- **Toxicology Testing:** Cell cultures are used to assess the toxicity of

chemicals, cosmetics, and pharmaceuticals. Testing on human or animal cells provides an ethical and efficient way to predict how these substances might affect living organisms, aiding regulatory bodies in determining safety standards.

- **Stem Cell Research and Regenerative Medicine:** Stem cells are cultured to study their potential for developing into various cell types. In regenerative medicine, cultured stem cells hold promise for repairing damaged tissues, offering potential treatments for injuries and degenerative diseases.

## 3.9 CHALLENGES IN CELL CULTURE

Despite their utility, cell culture techniques face several challenges. One major issue is the risk of contamination, which can affect the reliability of results and the health of cultures. Additionally, maintaining consistent cell behavior across passages can be difficult, as cells may undergo genetic and epigenetic changes over time. These changes can affect their growth characteristics and experimental outcomes, making it essential to monitor cultures closely. Another challenge is the discrepancy between in vitro (cell culture) and in vivo (within an organism) environments, as cells grown in artificial conditions may behave differently from those within the body.

## 3.10 FUTURE DIRECTIONS IN CELL CULTURE

Advancements in cell culture technology are continually improving the accuracy and applicability of in vitro studies. Organoid culture, for example, allows researchers to grow miniature, organ-like structures from stem cells, providing a more accurate model of human physiology. Additionally, 3D cell culture techniques are advancing, allowing for complex models that closely resemble tissues in their natural state. These developments in cell culture methods offer exciting possibilities for drug development, disease modeling, and personalized medicine.

## 3.11 CONCLUSION

Cell culture techniques are foundational in life sciences, providing powerful tools to study cellular behaviors, test drugs, and explore genetic mechanisms. From primary cultures to immortalized cell lines, researchers have access to a variety of culture types and methods that allow them to investigate diverse biological questions. Though challenges such as contamination and maintaining cell viability persist, advancements in cell culture technology are continuously enhancing the scope and reliability of these techniques. As cell culture methods continue to evolve, they will undoubtedly play a vital role in future discoveries, especially in fields like regenerative medicine, cancer research, and personalized drug testing.

# CHAPTER 4

# RECOMBINANT DNA TECHNIQUES AND MOLECULAR CLONING

Recombinant DNA technology and molecular cloning are cornerstones of modern molecular biology. These techniques enable scientists to manipulate genetic material, allowing for the study and modification of genes in ways that were previously impossible. They form the basis for advancements in genetic engineering, medicine, agriculture, and biotechnology. Recombinant DNA technology involves combining DNA from different sources to create new genetic combinations, while molecular cloning is the process of making copies of a specific DNA segment. Together, these techniques allow researchers to analyze gene function, produce therapeutic proteins, develop genetically modified organisms, and create vaccines.

## 4.1 INTRODUCTION TO RECOMBINANT DNA AND MOLECULAR CLONING

Recombinant DNA technology was first developed in the early 1970s, revolutionizing molecular biology and genetics. Recombinant DNA is artificially created by combining DNA sequences from two or more organisms, creating a "recombinant" molecule with novel properties. This technique laid the foundation for molecular cloning, which involves copying specific DNA sequences and studying them in detail. Molecular cloning allows researchers to isolate, amplify, and analyze genes, providing a crucial tool for genetic research, biotechnology, and medical advancements. Together, recombinant DNA technology and molecular cloning enable scientists to manipulate genetic material to better understand the functions of genes and create genetically modified organisms with desirable traits.

## 4.2 KEY PRINCIPLES OF RECOMBINANT DNA TECHNOLOGY

Recombinant DNA technology operates on several fundamental principles, which guide the manipulation and combination of DNA from different sources.

### 4.2.1 Restriction Enzyme

A restriction enzyme, restriction endonuclease, REase, ENase or *restrictase* is an enzyme that cleaves DNA into fragments at or near specific recognition sites within molecules known as restriction sites. Restriction enzymes are one class of the broader endonuclease group of enzymes. Restriction enzymes are commonly classified into five types, which differ in their structure and whether they cut their DNA substrate at their recognition site, or if the recognition and cleavage sites are separate from one another. To cut DNA, all restriction enzymes make two incisions, once through each sugar-phosphate backbone (i.e. each strand) of the DNA double helix.

These enzymes are found in bacteria and archaea and provide a defense mechanism against invading viruses. Inside a prokaryote, the restriction enzymes selectively cut up *foreign* DNA in a process called *restriction digestion*; meanwhile, host DNA is protected by a modification enzyme (a methyltransferase) that modifies the prokaryotic DNA and blocks cleavage. Together, these two processes form the restriction modification system.

More than 3,600 restriction endonucleases are known which represent over 250 different specificities. Over 3,000 of these have been studied in detail, and more than 800 of these are available commercially. These enzymes are routinely used for DNA modification in laboratories, and they are a vital tool in molecular cloning.

## History

The term restriction enzyme originated from the studies of phage λ, a virus that infects bacteria, and the phenomenon of host-controlled restriction and modification of such bacterial phage or bacteriophage. The phenomenon was first identified in work done in the laboratories of Salvador Luria, Jean Weigle and Giuseppe Bertani in the early 1950s. It was found that, for a bacteriophage λ that can grow well in one strain of *Escherichia coli*, for example *E. coli* C, when grown in another strain, for example *E. coli* K, its yields can drop significantly, by as much as three to five orders of magnitude. The host cell, in this example *E. coli* K, is known as the restricting host and appears to have the ability to reduce the biological activity of the phage λ. If a phage becomes established in one strain, the ability of that phage to grow also becomes restricted in other strains. In the 1960s, it was shown in work done in the laboratories of Werner Arber and Matthew Meselson that the restriction is caused by an enzymatic cleavage of the phage DNA, and the enzyme involved was therefore termed a restriction enzyme.

The restriction enzymes studied by Arber and Meselson were type I restriction enzymes, which cleave DNA randomly away from the recognition site. In 1970, Hamilton O. Smith, Thomas Kelly and Kent Wilcox isolated and characterized the first type II restriction enzyme, HindII, from the bacterium *Haemophilus influenzae*. Restriction enzymes of this type are more useful for laboratory work as they cleave DNA at the site of their recognition sequence and are the most commonly used as a molecular biology tool. Later, Daniel Nathans and Kathleen Danna showed that cleavage of simian virus 40 (SV40) DNA by restriction enzymes yields specific fragments that can be separated using polyacrylamide gel electrophoresis, thus showing that restriction enzymes can also be used for mapping DNA. For their work in the discovery and characterization of restriction enzymes, the 1978 Nobel Prize for Physiology or Medicine was awarded to Werner Arber, Daniel Nathans, and Hamilton O. Smith. The discovery of restriction enzymes allows DNA to be manipulated, leading to the development of recombinant DNA technology that has many applications, for example, allowing the large scale production of proteins such as human insulin used by diabetic patients.

## Origins

Restriction enzymes likely evolved from a common ancestor and became widespread via horizontal gene transfer. In addition, there is mounting evidence that restriction endonucleases evolved as a selfish genetic element.

## Recognition Site

Restriction enzymes recognize a specific sequence of nucleotides and produce a double-stranded cut in the DNA. The recognition sequences can also be classified by the number of bases in its recognition site, usually between 4 and 8 bases, and the number of bases in the sequence will determine how often the site will appear by chance in any given genome, e.g., a 4-base pair sequence would theoretically occur once every 4^4 or 256bp, 6 bases, 4^6 or 4,096bp, and 8 bases would be 4^8 or 65,536bp. Many of them are palindromic, meaning the base sequence reads the same backwards and forwards. In theory, there are two types of palindromic sequences that can be possible in DNA. The *mirror-like* palindrome is similar to those found in ordinary text, in which a sequence reads the same forward and backward on a single strand of DNA, as in GTAATG. The *inverted repeat* palindrome is also a sequence that reads the same forward and backward, but the forward and backward sequences are found in complementary DNA strands (i.e., of double-stranded DNA), as in GTATAC (GTATAC being complementary to CATATG). Inverted repeat palindromes are more common and have greater biological importance than mirror-like palindromes.

EcoRI digestion produces "sticky" ends,

G|AATTC
CTTAA|G

whereas SmaI restriction enzyme cleavage produces "blunt" ends:

CCC|GGG
GGG|CCC

Recognition sequences in DNA differ for each restriction enzyme, producing differences in the length, sequence and strand orientation (5' end or 3' end) of a sticky-end "overhang" of an enzyme restriction.

Different restriction enzymes that recognize the same sequence are known as neoschizomers. These often cleave in different locales of the sequence. Different enzymes that recognize and cleave in the same location are known as isoschizomers.

## Types

Naturally occurring restriction endonucleases are categorized into five groups (Types I, II, III, IV, and V) based on their composition and enzyme cofactor requirements, the nature of their target sequence, and the position of their DNA cleavage site relative to the target sequence. DNA sequence analysis of restriction enzymes however show great variations, indicating that there are more than four types. All types of enzymes recognize specific short DNA sequences and carry out the endonucleolytic cleavage of DNA to give specific fragments with terminal 5'-phosphates. They differ in their recognition sequence, subunit composition, cleavage position, and cofactor requirements:

- **Type I Enzymes (EC 3.1.21.3)** cleave at sites remote from a recognition site; require both ATP and S-adenosyl-L-methionine to function; multifunctional protein with both restriction digestion and methylase (EC 2.1.1.72) activities.
- **Type II Enzymes (EC 3.1.21.4)** cleave within or at short specific distances from a recognition site; most require magnesium; single function (restriction digestion) enzymes independent of methylase.
- **Type III Enzymes (EC 3.1.21.5)** cleave at sites a short distance from a recognition site; require ATP (but do not hydrolyse it); S-adenosyl-L-methionine stimulates the reaction but is not required; exist as part of a complex with a modification methylase (EC 2.1.1.72).
- **Type IV Enzymes** target modified DNA, e.g., methylated, hydroxymethylated and glucosyl-hydroxymethylated DNA
- **Type V Enzymes** utilize guide RNAs (gRNAs)

### Type I

Type I restriction enzymes were the first to be identified and were first identified in two different strains (K-12 and B) of *E. coli*. These enzymes cut at a site that differs, and is a random distance (at least 1000 bp) away, from their recognition site. Cleavage at these random sites follows a process of DNA translocation, which shows that these enzymes are also molecular motors. The recognition site is asymmetrical and is composed of two specific portions—one containing 3–4 nucleotides, and another containing 4–5 nucleotides—separated by a non-specific spacer of about 6–8 nucleotides. These enzymes are multifunctional and are capable of both restriction digestion and modification activities, depending upon the methylation status of the target DNA. The cofactors S-Adenosyl methionine (AdoMet), hydrolyzed adenosine triphosphate (ATP), and magnesium ($Mg^{2+}$) ions, are required for their full activity. Type I restriction enzymes possess three subunits called HsdR, HsdM, and HsdS; HsdR is required for restriction digestion; HsdM is necessary for adding methyl groups to host DNA (methyltransferase activity), and HsdS is important for specificity of the recognition (DNA-binding) site in addition to both restriction digestion (DNA cleavage) and modification (DNA methyltransferase) activity.

### Type II

Typical type II restriction enzymes differ from type I restriction enzymes in several ways. They form homodimers, with recognition sites that are usually undivided and palindromic and 4–8 nucleotides in length. They recognize and cleave DNA at the same site, and they do not use ATP or AdoMet for their activity—they usually require only $Mg^{2+}$ as a cofactor. These enzymes cleave the phosphodiester bond of double helix DNA. It can either cleave at the center of both strands to yield a blunt end, or at a staggered position leaving overhangs called sticky ends. These are the most commonly available and used restriction enzymes. In the 1990s and early 2000s, new enzymes from this family were discovered that did not follow all the classical criteria of this enzyme class, and new subfamily nomenclature was developed to divide this large family into subcategories based on deviations from typical characteristics of type II enzymes. These subgroups are defined using a letter suffix.

Type IIB restriction enzymes (e.g., BcgI and BplI) are multimers, containing more than one subunit. They cleave DNA on both sides of their recognition to cut out the recognition site. They require both AdoMet and $Mg^{2+}$ cofactors. Type IIE restriction endonucleases (e.g., NaeI) cleave DNA following interaction with two copies of their recognition sequence. One recognition site acts as the target for cleavage, while the other acts as an allosteric effector that speeds up or improves the efficiency of enzyme cleavage.

Similar to type IIE enzymes, type IIF restriction endonucleases (e.g. NgoMIV) interact with two copies of their recognition sequence but cleave both sequences at the same time. Type IIG restriction endonucleases (e.g., RM.Eco57I) do have a single subunit, like classical Type II restriction enzymes, but require the cofactor AdoMet to be active. Type IIM restriction endonucleases, such as DpnI, are able to recognize and cut methylated DNA. Type IIS restriction endonucleases (e.g. FokI) cleave DNA at a defined distance from their non-palindromic asymmetric recognition sites; this characteristic is widely used to perform in-vitro cloning techniques such as Golden Gate cloning. These enzymes may function as dimers. Similarly, Type IIT restriction enzymes (e.g., Bpu10I and BslI) are composed of two different subunits.

Some recognize palindromic sequences while others have asymmetric recognition sites.

### Type III

Type III restriction enzymes (e.g., EcoP15) recognize two separate non-palindromic sequences that are inversely oriented. They cut DNA about 20–30 base pairs after the recognition site. These enzymes contain more than one subunit and require AdoMet and ATP cofactors for their roles in DNA methylation and restriction digestion, respectively. They are components of prokaryotic DNA restriction-modification mechanisms that protect the organism against invading foreign DNA.

Type III enzymes are hetero-oligomeric, multifunctional proteins composed of two subunits, Res (P08764) and Mod (P08763). The Mod subunit recognises the DNA sequence specific for the system and is a modification methyltransferase; as such, it is functionally equivalent to the M and S subunits of type I restriction endonuclease. Res is required for restriction digestion, although it has no enzymatic activity on its own.

Type III enzymes recognise short 5–6 bp-long asymmetric DNA sequences and cleave 25–27 bp downstream to leave short, single-stranded 5' protrusions. They require the presence of two inversely oriented unmethylated recognition sites for restriction digestion to occur. These enzymes methylate only one strand of the DNA, at the N-6 position of adenine residues, so newly replicated DNA will have only one strand methylated, which is sufficient to protect against restriction digestion. Type III enzymes belong to the beta-subfamily of N6 adenine methyltransferases, containing the nine motifs that characterise this family, including motif I, the AdoMet binding pocket (FXGXG), and motif IV, the catalytic region (S/D/N (PP) Y/F).

### Type IV

Type IV enzymes recognize modified, typically methylated DNA and are exemplified by the McrBC and Mrr systems of *E. coli*.

### Type V

Type V restriction enzymes (e.g., the cas9-gRNA complex from CRISPRs) utilize guide RNAs to target specific non-palindromic sequences found on invading organisms. They can cut DNA of variable length, provided that a suitable guide RNA is provided. The flexibility and ease of use of these enzymes make them promising for future genetic engineering applications.

### Artificial Restriction Enzymes

Artificial restriction enzymes can be generated by fusing a natural or engineered DNA-binding domain to a nuclease domain (often the cleavage domain of the type

IIS restriction enzyme FokI). Such artificial restriction enzymes can target large DNA sites (up to 36 bp) and can be engineered to bind to desired DNA sequences. Zinc finger nucleases are the most commonly used artificial restriction enzymes and are generally used in genetic engineering applications, but can also be used for more standard gene cloning applications. Other artificial restriction enzymes are based on the DNA binding domain of TAL effectors.

In 2013, a new technology CRISPR-Cas9, based on a prokaryotic viral defense system, was engineered for editing the genome, and it was quickly adopted in laboratories.

In 2017, a group from University of Illinois reported using an Argonaute protein taken from *Pyrococcus furiosus* (PfAgo) along with guide DNA to edit DNA *in vitro* as artificial restriction enzymes.

Artificial ribonucleases that act as restriction enzymes for RNA have also been developed. A PNA-based system, called a PNAzyme, has a Cu(II)-2,9-dimethylphenanthroline group that mimics ribonucleases for specific RNA sequence and cleaves at a non-base-paired region (RNA bulge) of the targeted RNA formed when the enzyme binds the RNA. This enzyme shows selectivity by cleaving only at one site that either does not have a mismatch or is kinetically preferred out of two possible cleavage sites.

## Nomenclature

Since their discovery in the 1970s, many restriction enzymes have been identified; for example, more than 3500 different Type II restriction enzymes have been characterized. Each enzyme is named after the bacterium from which it was isolated, using a naming system based on bacterial genus, species and strain. For example, the name of the EcoRI restriction enzyme was derived as shown in the Table.

**Table 1: Derivation of the EcoRI name**

| Abbreviation | Meaning | Description |
|---|---|---|
| E | *Escherichia* | genus |
| co | *coli* | specific species |
| R | RY13 | strain |
| I | First identified | order of identification in the bacterium |

## Applications

Isolated restriction enzymes are used to manipulate DNA for different scientific applications.

They are used to assist insertion of genes into plasmid vectors during gene cloning and protein production experiments. For optimal use, plasmids that are commonly used for gene cloning are modified to include a short polylinker sequence (called the multiple cloning site, or MCS) rich in restriction recognition sequences. This allows flexibility when inserting gene fragments into the plasmid vector; restriction sites contained naturally within genes influence the choice of endonuclease for digesting the DNA, since it is necessary to avoid restriction of wanted DNA while intentionally cutting the ends of the DNA. To clone a gene fragment into a vector, both plasmid DNA and gene insert are typically cut with the same restriction enzymes, and then glued together with the assistance of an enzyme known as a DNA ligase.

Restriction enzymes can also be used to distinguish gene alleles by specifically recognizing single base changes in DNA known as single-nucleotide polymorphisms (SNPs). This is however only possible if a SNP alters the restriction site present in the allele. In this method, the restriction enzyme can be used to genotype a DNA sample without the need for expensive gene sequencing. The sample is first digested with the restriction enzyme to generate DNA fragments, and then the different sized fragments separated by gel electrophoresis. In general, alleles with correct restriction sites will generate two visible bands of DNA on the gel, and those with altered restriction sites will not be cut and will generate only a single band. A DNA map by restriction digest can also be generated that can give the relative positions of the genes. The different lengths of DNA generated by restriction digest also produce a specific pattern of bands after gel electrophoresis, and can be used for DNA fingerprinting.

In a similar manner, restriction enzymes are used to digest genomic DNA for gene analysis by Southern blot. This technique allows researchers to identify how many copies (or paralogues) of a gene are present in the genome of one individual, or how many gene mutations (polymorphisms) have occurred within a population. The latter example is called restriction fragment length polymorphism (RFLP).

Artificial restriction enzymes created by linking the FokI DNA cleavage domain with an array of DNA binding proteins or zinc finger arrays, denoted zinc finger nucleases (ZFN), are a powerful tool for host genome editing due to their enhanced sequence specificity. ZFN work in pairs, their dimerization being mediated in-situ through the FokI domain. Each zinc finger array (ZFA) is capable of recognizing 9–12 base pairs, making for 18–24 for the pair. A 5–7 bp spacer between the cleavage sites further enhances the specificity of ZFN, making them a safe and more precise tool that can be applied in humans. A recent Phase I clinical trial of ZFN for the targeted abolition of the CCR5 co-receptor for HIV-1 has been undertaken.

Others have proposed using the bacteria R-M system as a model for devising human anti-viral gene or genomic vaccines and therapies since the RM system serves an innate defense-role in bacteria by restricting tropism by bacteriophages.

There is research on REases and ZFN that can cleave the DNA of various human viruses, including HSV-2, high-risk HPVs and HIV-1, with the ultimate goal of inducing target mutagenesis and aberrations of human-infecting viruses. The human genome already contains remnants of retroviral genomes that have been inactivated and harnessed for self-gain. Indeed, the mechanisms for silencing active L1 genomic retroelements by the three prime repair exonuclease 1 (TREX1) and excision repair cross complementing 1(ERCC) appear to mimic the action of RM-systems in bacteria, and the non-homologous end-joining (NHEJ) that follows the use of ZFN without a repair template.

### 4.2.2 Ligation

Ligation is the joining of two nucleotides, or two nucleic acid fragments, into a single polymeric chain through the action of an enzyme known as a ligase. The reaction involves the formation of a phosphodiester bond between the 3'-hydroxyl terminus of one nucleotide and the 5'-phosphoryl terminus of another nucleotide, which results in the two nucleotides being linked consecutively on a single strand. Ligation works in fundamentally the same way for both DNA and RNA. A cofactor is generally involved in the reaction, usually ATP or $NAD^{+}$. Eukaryotic ligases belong to the ATP type, while the NAD+ type are found in bacteria (e.g. *E. coli*).

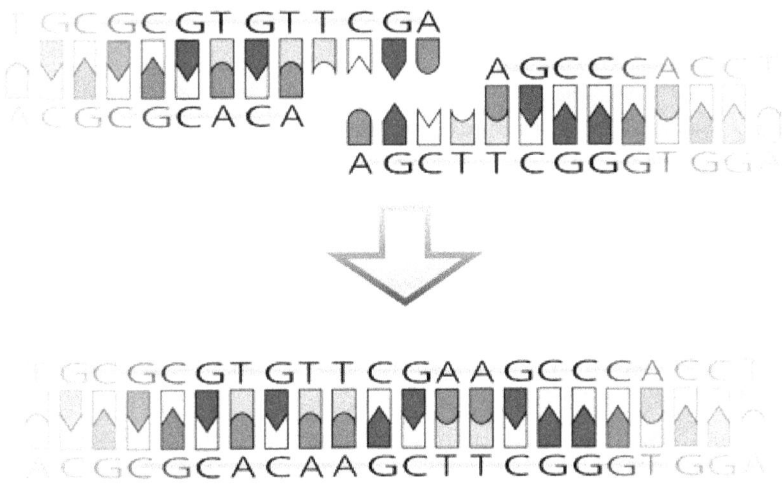


**Fig 1: A Sticky End Ligation**

Ligation occurs naturally as part of numerous cellular processes, including DNA replication, transcription, splicing, and recombination, and is also an essential laboratory procedure in molecular cloning, whereby DNA fragments are joined to create recombinant DNA molecules (such as when a foreign DNA fragment is inserted into a plasmid). The discovery of DNA ligase dates back to 1967 and was an important event in the field of molecular biology. Ligation in the laboratory is normally performed using T4 DNA ligase. It is broadly used *in vitro* due to its capability of joining sticky-ended fragments as well as blunt-ended fragments. However, procedures for ligation without the use of standard DNA ligase are also popular. Human DNA ligase abnormalities have been linked to pathological disorders characterized by immunodeficiency, radiation sensitivity, and developmental problems.

## Ligation Reaction

The mechanism of the ligation reaction was first elucidated in the laboratory of I. Robert Lehman. Two fragments of DNA may be joined by DNA ligase which catalyzes the formation of a phosphodiester bond between the 3'-hydroxyl group (-OH) at one end of a strand of DNA and the 5'-phosphate group (-PO4) of another. In animals and bacteriophages, ATP is used as the energy source for the ligation, while in bacteria, $NAD^+$ is used.

The DNA ligase first reacts with ATP or $NAD^+$, forming a ligase-AMP intermediate with the AMP linked to the ε-amino group of lysine in the active site of the ligase via a phosphoramide bond. This adenylyl group is then transferred to the phosphate group at the 5' end of a DNA chain, forming a DNA-adenylate complex. Finally, a phosphodiester bond between the two DNA ends is formed via the nucleophilic attack of the 3'-hydroxyl at the end of a DNA strand on the activated 5′-phosphoryl group of another.

A nick in the DNA (i.e., a break in one strand of a double-stranded DNA) can be repaired very efficiently by the ligase. However, a complicating feature of ligation presents itself when ligating two separate DNA ends as the two ends need to come together before the ligation reaction can proceed. In the ligation of DNA with sticky or cohesive ends, the protruding strands of DNA may be annealed together already, therefore it is a relatively efficient process as it is equivalent to repairing two nicks in the DNA. However, in the ligation of blunt-ends, which lack protruding ends for the DNA to anneal together, the process is dependent on random collision for the ends to align together before they can be ligated, and is consequently a much less efficient process. The DNA ligase from *E. coli* cannot ligate blunt-ended DNA except under conditions of molecular crowding, and it is therefore not normally used for ligation in the laboratory. Instead the DNA ligase from phage T4 is used as it can ligate blunt-ended DNA as well as single-stranded DNA.

## Factors Affecting Ligation

Factors that affect an enzyme-mediated chemical reaction would naturally affect a ligation reaction, these include the concentration of enzyme and the reactants, the temperature of reaction and the length of time of incubation. Ligation is complicated by the fact that the reaction can involve both inter- and intramolecular reactions, but the desired ligation products in many ligation reactions (e.g. ligating a DNA fragment into a vector) needed first to be inter-molecular, i.e. between two different DNA molecules, followed by an intra-molecular reaction to seal and circularize the molecule. For efficient ligation, an additional annealing step is also necessary.

The three steps to form a new phosphodiester bond during ligation are: enzyme adenylylation, adenylyl transfer to DNA, and nick sealing. Mg(2+) is a cofactor for catalysis, therefore at high concentration of Mg(2+) the ligation efficiency is high. If the concentration of Mg(2+) is limited, the nick- sealing is the rate- limiting reaction of the process, and adenylylated DNA intermediate stays in the solution. Such adenylylation of the enzyme restrains the rebinding to the adenylylated DNA intermediate comparison of an Achilles' heel of LIG1, and represents a risk if they are not fixed.

### DNA Concentration

The concentration of DNA can affect the rate of ligation, and whether the ligation is an inter-molecular or intra-molecular reaction. Ligation involves joining up the ends of a DNA with other ends, however, each DNA fragment has two ends, and if the ends are compatible, a DNA molecule can circularize by joining its own ends.

At high DNA concentration, there is a greater chance of one end of a DNA molecule meeting the end of another DNA, thereby forming intermolecular ligation. At a lower DNA concentration, the chance that one end of a DNA molecule would meet the other end of the same molecule increases, therefore intramolecular reaction that circularizes the DNA is more likely. The transformation efficiency of linear DNA is also much lower than circular DNA, and for the DNA to circularize, the DNA concentration should not be too high. As a general rule, the total DNA concentration should be less than 10 μg/ml.

The relative concentration of the DNA fragments, their length, as well as buffer conditions are also factors that can affect whether intermolecular or intramolecular reactions are favored.

The concentration of DNA can be artificially increased by adding condensing agents such as cobalt hexamine and biogenic polyamines such as spermidine, or by using crowding agents such as polyethylene glycol (PEG) which also increase the effective concentration of enzymes. Note however that additives such as cobalt hexamine can produce exclusively intermolecular reaction, resulting in linear concatemers rather than the circular DNA more suitable for transformation of plasmid DNA, and is therefore undesirable for plasmid ligation. If it is necessary to use additives in plasmid ligation, the use of PEG is preferable as it can promote intramolecular as well as intermolecular ligation.

### Ligase Concentration

As is usual for an enzyme, the higher the ligase concentration, the faster is the rate of ligation. Blunt-end ligation is much less efficient than sticky end ligation, so a higher concentration of ligase is used in blunt-end ligations. High DNA ligase concentration may be used in conjunction with PEG for a faster ligation, and they are the components often found in commercial kits designed for rapid ligation.

### Temperature

Two issues are involved when considering the temperature of a ligation reaction. First, the optimum temperature for DNA ligase activity which is 37°C, and second, the melting temperature ($T_m$) of the DNA ends to be ligated. The melting temperature is dependent on length and base composition of the DNA overhang—the greater the number of G and C, the higher the $T_m$ since there are three hydrogen bonds formed between G-C base pair compared to two for A-T base pair—with some contribution from the stacking of the bases between fragments. For the ligation reaction to proceed efficiently, the ends should be stably annealed, and in ligation experiments, the $T_m$ of the DNA ends is generally much lower than 37°C. The optimal temperature for ligating cohesive ends is therefore a compromise between the best temperature for DNA ligase activity and the $T_m$ where the ends can associate.

However, different restriction enzymes generates different ends, and the base composition of the ends produced by these enzymes may also differ, the melting temperature and therefore the optimal temperature can vary widely depending on the restriction enzymes used, and the optimum temperature for ligation may be between 4-15°C depending on the ends. Ligations also often involve ligating ends generated from different restriction enzymes in the same reaction mixture, therefore it may not be practical to select optimal temperature for a particular ligation reaction and most protocols simply choose 12-16°C, room temperature, or 4°C. When conducting a ligation at 4°C, it is necessary to increase the time of ligation reaction, for example by leaving the ligation mixture overnight or longer in the fridge.

### Buffer Composition

The ionic strength of the buffer used can affect the ligation. The kinds of cations presence can also influence the ligation reaction, for example, excess amount of $Na^+$ can cause the DNA to become more rigid and increase the likelihood of intermolecular ligation. At high concentration of monovalent cation (>200 mM) ligation can also be almost completely inhibited. The standard buffer used for ligation is designed to minimize ionic effects.

### Sticky-End Ligation

Restriction enzymes can generate a wide variety of ends in the DNA they digest, but in cloning experiments most commonly-used restriction enzymes generate a 4-base single-stranded overhang called the sticky or cohesive end (exceptions include *Nde*I which generates a 2-base overhang, and those that generate blunt ends). These sticky ends can anneal to other compatible ends and become ligated in a sticky-end (or cohesive end) ligation. *Eco*RI for example generates an AATT end, and since A and T have lower melting temperature than C and G, its melting temperature $T_m$ is low at around 6°C. For most restriction enzymes, the overhangs generated have a $T_m$ that is around 15°C. For practical purposes, sticky end ligations are performed at 12-16°C, or at room temperature, or alternatively at 4°C for a longer period.

For the insertion of a DNA fragment into a plasmid vector, it is preferable to use two different restriction enzymes to digest the DNA so that different ends are generated. The two different ends can prevent the religation of the vector without any insert, and it also allows the fragment to be inserted in a directional manner.

When it is not possible to use two different sites, then the vector DNA may need to be dephosphorylated to avoid a high background of recircularized vector DNA with no insert. Without a phosphate group at the ends the vector cannot ligate to itself, but can be ligated to an insert with a phosphate group. Dephosphorylation is commonly done using calf-intestinal alkaline phosphatase (CIAP) which removes the phosphate group from the 5′ end of digested DNA, but note that CIAP is not easy to inactivate and can interfere with ligation without an additional step to remove the CIAP, thereby resulting in failure of ligation. CIAP should not be used in excessive amount and should only be used when necessary. Shrimp alkaline phosphatase (SAP) or Antarctic phosphatase (AP) are suitable alternative as they can be easily inactivated.

## Blunt-End Ligation

Blunt end ligation does not involve base-pairing of the protruding ends, so any blunt end may be ligated to another blunt end. Blunt ends may be generated by restriction enzymes such as *Sma*I and *Eco*RV.

A major advantage of blunt-end cloning is that the desired insert does not require any restriction sites in its sequence as blunt-ends are usually generated in a PCR, and the PCR generated blunt-ended DNA fragment may then be ligated into a blunt-ended vector generated from restriction digest.

Blunt-end ligation, however, is much less efficient than sticky end ligation, typically the reaction is 100X slower than sticky-end ligation. Since blunt-end does not have protruding ends, the ligation reaction depends on random collisions between the blunt-ends and is consequently much less efficient. To compensate for the lower efficiency, the concentration of ligase used is higher than sticky end ligation (10x or more). The concentration of DNA used in blunt-end ligation is also higher to increase the likelihood of collisions between ends, and longer incubation time may also be used for blunt-end ligations.

If both ends needed to be ligated into a vector are blunt-ended, then the vector needs to be dephosphorylated to minimize self-ligation. This may be done using CIAP, but caution in its use is necessary as noted previously. Since the vector has been dephosphorylated, and ligation requires the presence of a 5'-phosphate, the insert must be phosphorylated. Blunt-ended PCR product normally lacks a 5'-phosphate, therefore it needs to be phosphorylated by treatment with T4 polynucleotide kinase.

Blunt-end ligation is also reversibly inhibited by high concentration of ATP. PCR usually generates blunt-ended PCR products, but note that PCR using *Taq* polymerase can add an extra adenine (A) to the 3' end of the PCR product. This property may be exploited in TA cloning where the ends of the PCR product can anneal to the T end of a vector. TA ligation is therefore a form of sticky end ligation. Blunt-ended vectors may be turned into vector for TA ligation with dideoxythymidine triphosphate (ddTTP) using terminal transferase.

## General Guidelines

For the cloning of an insert into a circular plasmid:

- The total DNA concentration used should be less than 10 μg/ml as the plasmid needs to recircularize.
- The molar ratio of insert to vector is usually used at around 3:1. Very high ratio may produce multiple inserts. The ratio may be adjusted depending on the size of the insert, and other ratios may be used, such as 1:1.

## Trouble-Shooting

Sometimes ligation fail to produce the desired ligated products, and some of the possible reasons may be:

- **Damaged DNA:** Over-exposure to UV radiation during preparation of DNA for ligation can damage the DNA and significantly reduce transformation efficiency. A higher-wavelength UV radiation (365 nm) which cause less damage to DNA should be used if it is necessary work for work on the DNA on a UV transilluminator for an extended period of time. Addition of cytidine or guanosine to the electrophoresis buffer at 1 mM concentration however may protect the DNA from damage.
- Incorrect usage of CIAP or its inefficient inactivation or removal.
- Excessive amount of DNA used.

- Incomplete DNA digest – The vector DNA that is incompletely digested will give rise to a high background, and this may be checked by doing a ligation without insert as a control. Insert that is not completely digested will also not ligate properly and circularize. When digesting a PCR product, make sure that sufficient extra bases have been added to the 5'-ends of the oligonucleotides used for PCR as many restriction enzymes require a minimum number of extra basepairs for efficient digest. The information on the minimum basepair required is available from restriction enzyme suppliers such as in the catalog of New England Biolabs.
- Incomplete ligation – Blunt-ends DNA (e.g. *Sma*I) and some sticky-ends DNA (e.g. *Nde*I) that have low-melting temperature require more ligase and longer incubation time.
- Protein expressed from ligated gene insert is toxic to cells.
- Homologous sequence in insert to sequence in plasmid DNA resulting in deletion.
- High concentration of EDTA or salts that acts as an inhibitors.

## Other Methods of DNA Ligation

A number of commercially available DNA cloning kits use other methods of ligation that do not require the use of the usual DNA ligases. These methods allow cloning to be done much more rapidly, as well as allowing for simpler transfer of cloned DNA insert to different vectors. These methods however require the use of specially designed vectors and components, and may lack flexibility.

### Topoisomerase-Mediated Ligation

Topoisomerase can be used instead of ligase for ligation, and the cloning may be done more rapidly without the need for restriction digest of the vector or insert. In this TOPO cloning method a linearized vector is activated by attaching topoisomerase I to its ends, and this "TOPO-activated" vector may then accept a PCR product by ligating to both of the 5' ends of the PCR product, the topoisomerase is released and a circular vector is formed in the process.

### Homologous Recombination

Different types of ligases found in the studied organisms. For instance, Nicotinamide adenine dinucleotide (NAD+)-dependent ligase was found and isolated from bacterial organism, known as *E. coli* in second third of 20th century. Since then, this model has been widely used to study that DNA ligase family. Moreover, it is found in all bacteria. Examples of genes present in *E. coli* are LigA, which has essential functions affecting bacterial growth, and LigB.

In mammals, including human 3 genes, namely Lig1, Lig3, Lig4 were identified. All eukaryotes contain multiple types of DNA ligases encoded by Lig genes. The smallest known eukaryotic ligase is *Chlorella* virus DNA ligase (ChVLig). It contains only 298 amino acids. When ChVLig is the only source of ligase in the cell, it can continue to support mitotic development, and nonhomologous end joining in budding yeasts. DNA Ligase I (Lig1) is accountable for Okazaki Fragments ligation. It is consist of 919 amino acids. In a complex process of DNA replication, DNA Ligase I recruited to replications machinery by protein interactions.

Lig1 plays role in cell division in plants and yeasts. Knockout of the Lig1 gene is lethal in yeasts and some plants sprouts. Nevertheless, studies of mouse embryogenesis have shown that until the middle of the growth process embryo developing without DNA ligase I.

Enzymatic ligation has been used in various studies related to DNA nanostructures and lead to increase of efficiency and stability. One of the methods is sealing of covalent DNA bond, namely phosphodiester bond and nicks. Reconstruction of those structures performed with assistance of ligation. For instance, T4 DNA ligase serve as a catalyst for sealing of a nick between 3 prime and 5 prime ends of DNA to make up strong phosphodiester bond. Ligated structures have higher thermal stability values.

T4 DNA ligase has many valuable properties such as already mentioned catalytic, but it is also responsible for sealing of the gaps between DNA strands, nick-closing activity, repair of the DNA damage, etc.

In nanostructures architecture, molecular biology researches - ssDNA is an important application model. T4 DNA ligase used to cyclize short ssDNA fragments, but process is complicated by formation of secondary structures. On the other hand, *Taq* DNA ligase is a thermostable enzyme which can be applied at higher temperatures (45, 55 and 65 °C respectively). Since at these temperature range secondary structures less stable it is enhance cyclization efficiency of oligonucleotides. The kinetic, biological, and other parameters of nanostructures are influenced by presence of the secondary structures in DNA rings. However, *Taq* DNA ligation occur only when two complementary DNA strands are perfectly paired and have no gaps in between.

Analysis of ligases activities, mutations, deficiencies widely used in drug design and biological researches to investigate diseases, pathologies developments and related rare acquired or inherited syndromes (e.g. DNA ligase IV syndrome).

The ligation procedure is prevalent in molecular biology cloning techniques, and it has been applied to define and characterize specific nucleotide sequences in the genome using Ligase Chain Reaction (LCR) or Polymerase Chain Reaction (PCR)-based amplification of ligated probes.

## Analysis

Ligation may also serve as a DNA analysis method. Some techniques employ rolling circle amplification. The most notable of these is described by Smolina *et al.*, 2007 & Smolina *et al.*, 2008 using fluorescence in situ hybridization and peptide nucleic acids. They developed and employed this technique for analyses of bacterial chromosomes.

### 4.2.3 Vector (Molecular Biology)

In molecular cloning, a vector is any particle (e.g., plasmids, cosmids, Lambda phages) used as a vehicle to artificially carry a foreign nucleic sequence – usually DNA – into another cell, where it can be replicated and/or expressed. A vector containing foreign DNA is termed recombinant DNA. The four major types of vectors are plasmids, viral vectors, cosmids, and artificial chromosomes. Of these, the most commonly used vectors are plasmids. Common to all engineered vectors are an origin of replication, a multicloning site, and a selectable marker.

The vector itself generally carries a DNA sequence that consists of an insert (in this case the transgene) and a larger sequence that serves as the "backbone" of the vector. The purpose of a vector which transfers genetic information to another cell is typically to isolate, multiply, or express the insert in the target cell.

All vectors may be used for cloning and are therefore cloning vectors, but there are also vectors designed specially for cloning, while others may be designed specifically for other purposes, such as transcription and protein expression. Vectors designed specifically for the expression of the transgene in the target cell are called expression vectors, and generally have a promoter sequence that drives expression of the transgene. Simpler vectors called transcription vectors are only capable of being transcribed but not translated: they can be replicated in a target cell but not expressed, unlike expression vectors. Transcription vectors are used to amplify their insert.

The manipulation of DNA is normally conducted on *E. coli* vectors, which contain elements necessary for their maintenance in *E. coli*. However, vectors may also have elements that allow them to be maintained in another organism such as yeast, plant or mammalian cells, and these vectors are called shuttle vectors.

Such vectors have bacterial or viral elements which may be transferred to the non-bacterial host organism, however other vectors termed intragenic vectors have also been developed to avoid the transfer of any genetic material from an alien species.

Insertion of a vector into the target cell is usually called transformation for bacterial cells, transfection for eukaryotic cells, although insertion of a viral vector is often called transduction.

## Characteristics

### Plasmids

Plasmids are double-stranded extra chromosomal and generally circular DNA sequences that are capable of replication using the host cell's replication machinery. Plasmid vectors minimalistically consist of an origin of replication that allows for semi-independent replication of the plasmid in the host. Plasmids are found widely in many bacteria, for example in *Escherichia coli*, but may also be found in a few eukaryotes, for example in yeast such as *Saccharomyces cerevisiae*. Bacterial plasmids may be conjugative/transmissible and non-conjugative:

- **Conjugative:** Mediate DNA transfer through conjugation and therefore spread rapidly among the bacterial cells of a population; e.g., F plasmid, many R and some col plasmids.
- **Nonconjugative:** Do not mediate DNA through conjugation, e.g., many R and col plasmids.

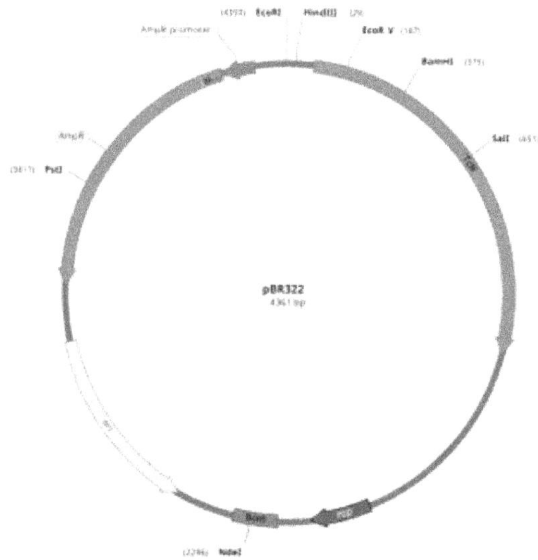


**Fig 2: The pBR322 Plasmid is One of the First Plasmids Widely Used as a Cloning Vector**

Plasmids with specially-constructed features are commonly used in laboratory for cloning purposes. These plasmid are generally non-conjugative but may have many more features, notably a "multiple cloning site" where multiple restriction enzyme cleavage sites allow for the insertion of a transgene insert. The bacteria containing the plasmids can generate millions of copies of the vector within the bacteria in hours, and the amplified vectors can be extracted from the bacteria for further manipulation. Plasmids may be used specifically as transcription vectors and such plasmids may lack crucial sequences for protein expression. Plasmids used for protein expression, called expression vectors, would include elements for translation of protein, such as a ribosome binding site, start and stop codons.

### Viral Vectors

Viral vectors are genetically engineered viruses carrying modified viral DNA or RNA that has been rendered noninfectious, but still contain viral promoters and the transgene, thus allowing for translation of the transgene through a viral promoter. However, because viral vectors frequently lack infectious sequences, they require helper viruses or packaging lines for large-scale transfection. Viral vectors are often designed to permanently incorporate the insert into the host genome, and thus leave distinct genetic markers in the host genome after incorporating the transgene. For example, retroviruses leaves a characteristic retroviral integration pattern after insertion that is detectable and indicates that the viral vector has incorporated into the host genome.

### Artificial Chromosomes

Artificial chromosomes are manufactured chromosomes in the context of yeast artificial chromosomes (YACs), bacterial artificial chromosomes (BACs), or human artificial chromosomes (HACs). An artificial chromosome can carry a much larger DNA fragment than other vectors. YACs and BACs can carry a DNA fragment up to 300,000 nucleotides long. Three structural necessities of an artificial chromosome include an origin of replication, a centromere, and telomeric end sequences.

## Transcription

Transcription of the cloned gene is a necessary component of the vector when expression of the gene is required: one gene may be amplified through transcription to generate multiple copies of mRNAs, the template on which protein may be produced through translation.

A larger number of mRNAs would express a greater amount of protein, and how many copies of mRNA are generated depends on the promoter used in the vector. The expression may be constitutive, meaning that the protein is produced constantly in the background, or it may be inducible whereby the protein is expressed only under certain condition, for example when a chemical inducer is added. These two different types of expression depend on the types of promoter and operator used.

Viral promoters are often used for constitutive expression in plasmids and in viral vectors because they normally force constant transcription in many cell lines and types reliably. Inducible expression depends on promoters that respond to the induction conditions: for example, the murine mammary tumor virus promoter only initiates transcription after dexamethasone application and the *Drosophila* heat shock promoter only initiates after high temperatures.

Some vectors are designed for transcription only, for example for *in vitro* mRNA production. These vectors are called transcription vectors. They may lack the sequences necessary for polyadenylation and termination, therefore may not be used for protein production.

## Expression

Expression vectors produce proteins through the transcription of the vector's insert followed by translation of the mRNA produced, they therefore require more components than the simpler transcription-only vectors. Expression in different host organism would require different elements, although they share similar requirements, for example a promoter for initiation of transcription, a ribosomal binding site for translation initiation, and termination signals.

### Prokaryotes Expression Vector

- **Promoter:** Commonly used inducible promoters are promoters derived from *lac* operon and the T7 promoter. Other strong promoters used include Trp promoter and Tac-Promoter, which are a hybrid of both the Trp and Lac Operon promoters.
- **Ribosome Binding Site (RBS):** Follows the promoter, and promotes efficient translation of the protein of interest.
- **Translation Initiation Site:** Shine-Dalgarno sequence enclosed in the RBS, 8 base-pairs upstream of the AUG start codon.

### Eukaryotes Expression Vector

Eukaryote expression vectors require sequences that encode for:

- **Polyadenylation Tail:** Creates a polyadenylation tail at the end of the transcribed pre-mRNA that protects the mRNA from exonucleases and ensures transcriptional and translational termination: stabilizes mRNA production.
- **Minimal UTR Length:** UTRs contain specific characteristics that may impede transcription or translation, and thus the shortest UTRs or none at all are encoded for in optimal expression vectors.
- **Kozak Sequence:** Vectors should encode for a Kozak sequence in the mRNA, which assembles the ribosome for translation of the mRNA.

## Features

Modern artificially-constructed vectors contain essential components found in all vectors, and may contain other additional features found only in some vectors:

- **Origin of Replication:** Necessary for the replication and maintenance of the vector in the host cell.

- **Promoter:** Promoters are used to drive the transcription of the vector's transgene as well as the other genes in the vector such as the antibiotic resistance gene. Some cloning vectors need not have a promoter for the cloned insert but it is an essential component of expression vectors so that the cloned product may be expressed.
- **Cloning Site:** This may be a multiple cloning site or other features that allow for the insertion of foreign DNA into the vector through ligation.
- **Genetic Markers:** Genetic markers for viral vectors allow for confirmation that the vector has integrated with the host genomic DNA.
- **Antibiotic Resistance:** Vectors with antibiotic-resistance open reading frames allow for survival of cells that have taken up the vector in growth media containing antibiotics through antibiotic selection.
- **Epitope:** Some vectors may contain a sequence for a specific epitope that can be incorporated into the expressed protein. It allows for antibody identification of cells expressing the target protein.
- **Reporter Genes:** Some vectors may contain a reporter gene that allow for identification of plasmid that contains inserted DNA sequence. An example is *lacZ-α* which codes for the N-terminus fragment of β-galactosidase, an enzyme that digests galactose. A multiple cloning site is located within *lacZ-α*, and an insert successfully ligated into the vector will disrupt the gene sequence, resulting in an inactive β-galactosidase. Cells containing vector with an insert may be identified using blue/white selection by growing cells in media containing an analogue of galactose (X-gal). Cells expressing β-galactosidase (therefore does not contain an insert) appear as blue colonies. White colonies would be selected as those that may contain an insert. Other commonly used reporters include green fluorescent protein and luciferase.
- **Targeting Sequence:** Expression vectors may include encoding for a targeting sequence in the finished protein that directs the expressed protein to a specific organelle in the cell or specific location such as the periplasmic space of bacteria.
- **Protein Purification Tags:** Some expression vectors include proteins or peptide sequences that allows for easier purification of the expressed protein. Examples include polyhistidine-tag, glutathione-S-transferase, and maltose binding protein. Some of these tags may also allow for increased solubility of the target protein. The target protein is fused to the protein tag, but a protease cleavage site positioned in the

polypeptide linker region between the protein and the tag allows the tag to be removed later.

## 4.3 MOLECULAR CLONING: THE PROCESS

Molecular cloning involves several steps, starting from DNA isolation to the final analysis of cloned DNA.

This process enables researchers to amplify and study specific genes or DNA fragments in detail.

- **Isolation of DNA:** The first step in molecular cloning is isolating the DNA of interest. This DNA can be derived from genomic DNA, cDNA (complementary DNA synthesized from mRNA), or synthetic DNA. The isolated DNA is then cut into fragments using restriction enzymes.
- **Insertion into a Vector:** After cutting the DNA into desired fragments, it is inserted into a vector that will carry the DNA into host cells. Plasmid vectors are widely used due to their versatility and ease of handling. The DNA and vector are mixed with DNA ligase, which joins them together to form a recombinant DNA molecule.
- **Transformation into Host Cells:** The recombinant DNA is introduced into a host cell, typically a bacterium like *E. coli*, through a process called transformation. In transformation, cells are made "competent" (more permeable to DNA) through chemical or electrical treatments, allowing the recombinant DNA to enter the cell.
- **Selection and Screening:** Once the host cells take up the recombinant DNA, they are grown on selective media containing antibiotics. Only the cells with the antibiotic resistance gene (contained within the vector) will survive, allowing researchers to identify cells that have successfully incorporated the recombinant DNA. Additional screening methods, such as blue-white screening or PCR, may be used to verify the presence of the desired DNA sequence.
- **Amplification and Analysis:** After selection, the recombinant DNA can be amplified within the host cells, producing multiple copies of the gene or DNA fragment. These cells can then be cultured to obtain a larger quantity of recombinant DNA, which can be extracted and analyzed through various techniques, including sequencing, gel electrophoresis, or expression studies.

## 4.4 APPLICATIONS OF RECOMBINANT DNA TECHNOLOGY AND MOLECULAR CLONING

Recombinant DNA technology and molecular cloning have a wide range of applications across multiple fields, including medicine, agriculture, environmental science, and basic research.

- **Production of Therapeutic Proteins**: Recombinant DNA technology allows for the production of therapeutic proteins, such as insulin, growth hormones, and clotting factors, by inserting human genes into bacterial cells. The bacteria express these proteins, which are then purified and used in medical treatments. This approach has transformed the production of essential proteins, making them more accessible and cost-effective.
- **Gene Therapy**: Gene therapy involves using recombinant DNA to introduce or correct faulty genes in patients with genetic disorders. By inserting therapeutic genes into a patient's cells, researchers hope to correct genetic defects and provide long-lasting treatments for conditions like cystic fibrosis, hemophilia, and muscular dystrophy.
- **Genetically Modified Organisms (GMOs)**: Recombinant DNA technology is used to create genetically modified organisms (GMOs) in agriculture. By introducing specific genes, scientists develop crops that are more resistant to pests, diseases, and environmental stress. GMOs can also be designed to increase nutritional content, providing potential solutions for food security and malnutrition.
- **Vaccine Development**: Recombinant DNA technology is used to create vaccines by inserting genes encoding viral or bacterial proteins into vectors. These recombinant proteins are then used to stimulate an immune response, providing protection against diseases like hepatitis B and HPV. This approach has also been critical in developing mRNA vaccines, which represent a novel application of molecular cloning techniques.
- **Basic Research and Functional Genomics**: Molecular cloning is a fundamental tool in studying gene function and regulation. By isolating and manipulating genes, scientists can examine how specific genes contribute to cellular processes, development, and disease. This research forms the basis for functional genomics, where large-scale analyses are conducted to understand the role of genes in biological systems.

## 4.5 ETHICAL CONSIDERATIONS AND CHALLENGES

Recombinant DNA technology and molecular cloning raise important ethical questions, especially regarding the modification of genetic material in humans and other organisms.

- **Safety of Genetic Engineering:** Concerns regarding genetically modified organisms (GMOs) focus on their potential environmental impact and their long-term effects on human health. Although many GMOs are considered safe, ongoing research and strict regulation are essential to address public concerns and ensure responsible use of this technology.
- **Gene Editing and Human Genetics:** Recombinant DNA technology, especially with the advent of CRISPR-Cas9, has made gene editing in humans a reality. While gene therapy holds potential for treating genetic disorders, the ability to edit genes also raises ethical concerns about designer babies and genetic enhancement. There is a growing debate on setting limits for genetic manipulation, particularly when it comes to human embryos.
- **Intellectual Property and Access:** The commercialization of recombinant DNA technology has led to debates about patenting genes and restricting access to life-saving treatments. Many argue that vital genetic technologies should be accessible to all, especially in developing countries where healthcare resources are limited.

## 4.6 CONCLUSION

Recombinant DNA techniques and molecular cloning have fundamentally transformed biological research, allowing scientists to explore, modify, and manipulate genes in unprecedented ways. From the production of therapeutic proteins and GMOs to the development of new vaccines, these technologies have widespread applications with the potential to address some of the world's most pressing challenges. Despite ethical concerns, responsible regulation and further research can help harness the power of recombinant DNA technology for the betterment of human health, agriculture, and environmental sustainability. As these techniques continue to evolve, they promise to shape the future of science, medicine, and society in profound ways.

# CHAPTER 5

# PREPARATIVE PROTEIN BIOCHEMISTRY

Preparative protein biochemistry is a critical field within the life sciences focused on the isolation, purification, and analysis of proteins to study their function, structure, and interactions. Proteins, as complex biomolecules, play fundamental roles in cellular processes, serving as enzymes, structural components, and signaling molecules. Understanding protein function and behavior requires carefully designed techniques to isolate and purify them from complex mixtures like cell lysates. Preparative protein biochemistry combines methods from chromatography, electrophoresis, and precipitation, providing essential tools for molecular biology, biochemistry, pharmacology, and biotechnology.

## 5.1 INTRODUCTION TO PREPARATIVE PROTEIN BIOCHEMISTRY

The field of preparative protein biochemistry involves techniques that allow for the extraction and purification of proteins in a form suitable for functional and structural analysis. Preparative methods are distinct from analytical approaches, which focus on identifying and quantifying proteins in small samples. Preparative techniques aim to yield larger quantities of pure protein, allowing for in-depth studies and functional assays. The process often begins with the selection of a protein source, followed by extraction, purification, and quality assessment. With advances in technology, preparative protein biochemistry has expanded to allow for precise manipulation and engineering of proteins, which has applications in drug development, diagnostics, and therapeutic research.

## 5.2 KEY PRINCIPLES OF PROTEIN EXTRACTION AND PURIFICATION

Successful protein purification requires careful attention to several key principles that preserve protein function and integrity while isolating the target protein.

- **Protein Stability**: Proteins can be sensitive to factors like temperature, pH, and salt concentration. Maintaining an optimal environment during extraction and purification is crucial for preserving protein activity and structure. Buffers are often used to maintain pH, while protease inhibitors can prevent protein degradation.

- **Specificity and Yield:** The goal of preparative protein biochemistry is to achieve a high yield of the desired protein with minimal contamination from other cellular components. Techniques such as affinity chromatography take advantage of specific interactions to bind the target protein, while other methods like ion-exchange chromatography separate proteins based on their charge.
- **Purity:** High purity is essential for downstream applications, as contaminants can interfere with biochemical analyses or structural studies. Purity can be assessed at each purification step using techniques like SDS-PAGE or Western blotting, ensuring that only the target protein is present.

## 5.3 PROTEIN EXTRACTION TECHNIQUES

Protein extraction is the first step in preparative protein biochemistry. This process involves breaking open cells to release proteins and separating them from other cellular components.

### 5.3.1 Cell Lysis

Lysis is the breaking down of the membrane of a cell, often by viral, enzymic, or osmotic mechanisms that compromise its integrity. A fluid containing the contents of lysed cells is called a *lysate*. In molecular biology, biochemistry, and cell biology laboratories, cell cultures may be subjected to lysis in the process of purifying their components, as in protein purification, DNA extraction, RNA extraction, or in purifying organelles.

Many species of bacteria are subject to lysis by the enzyme lysozyme, found in animal saliva, egg white, and other secretions. Phage lytic enzymes (lysins) produced during bacteriophage infection are responsible for the ability of these viruses to lyse bacterial cells.

Penicillin and related β-lactam antibiotics cause the death of bacteria through enzyme-mediated lysis that occurs after the drug causes the bacterium to form a defective cell wall. If the cell wall is completely lost and the penicillin was used on gram-positive bacteria, then the bacterium is referred to as a protoplast, but if penicillin was used on gram-negative bacteria, then it is called a spheroplast.

- **Cytolysis:** Cytolysis occurs when a cell bursts due to an osmotic imbalance that has caused excess water to move into the cell. Cytolysis can be prevented by several different mechanisms, including the contractile vacuole that exists in some paramecia, which rapidly pump water out of the cell. Cytolysis does not occur under normal conditions in plant cells because plant cells have a strong cell wall that contains the osmotic pressure, or turgor pressure, that would otherwise cause cytolysis to occur.

- **Oncolysis:** Oncolysis is the destruction of neoplastic cells or of a tumour. The term is also used to refer to the reduction of any swelling.
- **Plasmolysis:** Plasmolysis is the contraction of cells within plants due to the loss of water through osmosis. In a hypertonic environment, the cell membrane peels off the cell wall and the vacuole collapses. These cells will eventually wilt and die unless the flow of water caused by osmosis can stop the contraction of the cell membrane.
- **Immune Response:** Erythrocytes' hemoglobin release free radicals in response to pathogens when lysed by them. This can damage the pathogens.

## Applications

Cell lysis is used in laboratories to break open cells and purify or further study their contents. Lysis in the laboratory may be affected by enzymes or detergents or other chaotropic agents. Mechanical disruption of cell membranes, as by repeated freezing and thawing, sonication, pressure, or filtration may also be referred to as lysis. Many laboratory experiments are sensitive to the choice of lysis mechanism; often it is desirable to avoid mechanical shear forces that would denature or degrade sensitive macromolecules, such as proteins and DNA, and different types of detergents can yield different results. The unprocessed solution immediately after lysis but before any further extraction steps is often referred to as a *crude lysate*.

For example, lysis is used in western and Southern blotting to analyze the composition of specific proteins, lipids, and nucleic acids individually or as complexes. Depending on the detergent used, either all or some membranes are lysed. For example, if only the cell membrane is lysed then gradient centrifugation can be used to collect certain organelles. Lysis is also used for protein purification, DNA extraction, and RNA extraction.

## Methods

- **Chemical Lysis:** This method uses chemical disruption. It is the most popular and simple approach. Chemical lysis chemically deteriorates/ solubilizes the proteins and lipids present within the membrane of targeted cells. Common lysis buffers contain sodium hydroxide (NaOH) and sodium dodecyl sulfate (SDS). Cell lysis is best done at a pH range of 11.5-12.5. Although simple, it is a slow process, taking anywhere from 6 to 12 hours.

- **Acoustic Lysis:** This method uses ultrasonic waves to generate areas of high and low pressure which causes cavitation and in turn, cell lysis. Though this method usually comes out clean, it fails to be cost effective and consistent.
- **Mechanical Lysis:** This method uses physical penetration to pierce or cut a cell membrane.
- **Enzymatic Lysis:** This method uses enzymes such as lysozyme or proteases to disintegrate the cell membrane.

### 5.3.2 Centrifugation

Centrifugation is a mechanical process which involves the use of the centrifugal force to separate particles from a solution according to their size, shape, density, medium viscosity and rotor speed. The denser components of the mixture migrate away from the axis of the centrifuge, while the less dense components of the mixture migrate towards the axis. Chemists and biologists may increase the effective gravitational force of the test tube so that the precipitate (pellet) will travel quickly and fully to the bottom of the tube.

There is a correlation between the size and density of a particle and the rate that the particle separates from a heterogeneous mixture, when the only force applied is that of gravity. The larger the size and the larger the density of the particles, the faster they separate from the mixture. By applying a larger effective gravitational force to the mixture, like a centrifuge does, the separation of the particles is accelerated. This is ideal in industrial and lab settings because particles that would naturally separate over a long period of time can be separated in much less time.

The rate of centrifugation is specified by the angular velocity usually expressed as revolutions per minute (RPM), or acceleration expressed as $g$. The conversion factor between RPM and $g$ depends on the radius of the centrifuge rotor. The particles' settling velocity in centrifugation is a function of their size and shape, centrifugal acceleration, the volume fraction of solids present, the density difference between the particle and the liquid, and the viscosity. The most common application is the separation of solid from highly concentrated suspensions, which is used in the treatment of sewage sludges for dewatering where less consistent sediment is produced.

The centrifugation method has a wide variety of industrial and laboratorial applications; not only is this process used to separate two miscible substances, but also to analyze the hydrodynamic properties of macromolecules. It is one of the most important and commonly used research methods in biochemistry, cell and molecular biology. In the chemical and food industries, special centrifuges can process a continuous stream of particle turning into separated liquid like plasma.

Centrifugation is also the most common method used for uranium enrichment, relying on the slight mass difference between atoms of U-238 and U-235 in uranium hexafluoride gas.

## History

By 1923 Theodor Svedberg and his student H. Rinde had successfully analyzed large-grained sols in terms of their gravitational sedimentation. Sols consist of a substance evenly distributed in another substance, also known as a colloid. However, smaller grained sols, such as those containing gold, could not be analyzed. To investigate this problem Svedberg developed an analytical centrifuge, equipped with a photographic absorption system, which would exert a much greater centrifugal effect. In addition, he developed the theory necessary to measure molecular weight. During this time, Svedberg's attention shifted from gold to proteins.

By 1900, it had been generally accepted that proteins were composed of amino acids; however, whether proteins were colloids or macromolecules was still under debate. One protein being investigated at the time was hemoglobin. It was determined to have 712 carbon, 1,130 hydrogen, 243 oxygen, two sulfur atoms, and at least one iron atom. This gave hemoglobin a resulting weight of approximately 16,000 dalton (Da) but it was uncertain whether this value was a multiple of one or four (dependent upon the number of iron atoms present).

Through a series of experiments utilizing the sedimentation equilibrium technique, two important observations were made: hemoglobin has a molecular weight of 68,000 Da, suggesting that there are four iron atoms present rather than one, and that, no matter where the hemoglobin was isolated from, it had exactly the same molecular weight. How something of such a large molecular mass could be consistently found, regardless of where it was sampled from in the body, was unprecedented and favored the idea that proteins are macromolecules rather than colloids. In order to investigate this phenomenon, a centrifuge with even higher speeds was needed, and thus the ultracentrifuge was created to apply the theory of sedimentation-diffusion. The same molecular mass was determined, and the presence of a spreading boundary suggested that it was a single compact particle. Further application of centrifugation showed that under different conditions the large homogeneous particles could be broken down into discrete subunits. The development of centrifugation was a great advance in experimental protein science.

Linderstorm-Lang, in 1937, discovered that density gradient tubes could be used for density measurements. He discovered this when working with potato yellow-dwarf virus. This method was also used in Meselson and Stahl's famous experiment in which they proved that DNA replication is semi-conservative by using different isotopes of nitrogen. They used density gradient centrifugation to determine which isotope or isotopes of nitrogen were present in the DNA after cycles of replication.

## Mathematical Formula

In a liquid suspension, many particles or cells will gradually fall to the bottom of the container due to gravity; however, the amount of time taken for such separations is not feasible. Other particles, which are very small, can not be isolated at all in solution until they are exposed to a high centrifugal force. As the suspension is rotated at a certain speed or revolutions per minute (RPM), the centrifugal force allows the particles to travel radially away from the rotation axis. The general formula for calculating the revolutions per minute (RPM) of a centrifuge is:

$$RPM = \sqrt{\frac{g}{r}}$$

where *g* represents the relative centrifugal force (RCF) and *r* the radius from the center of the rotor to a point in the sample.

However, depending on the centrifuge model used, the respective angle of the rotor and the radius may vary, thus the formula gets modified. For example, the Sorvall #SS-34 rotor has a maximum radius of 10.8 cm, so the formula becomes

$$RPM = 299\sqrt{\frac{g}{r}}$$

which can further simplify to

$$RPM = 91\sqrt{g}$$

When compared to gravity, the particle force is called the 'Relative Centrifugal Force' (RCF). It is the perpendicular force exerted on the contents of the rotor as a result of the rotation, always relative to the gravity of the Earth, which measures the strength of rotors of different types and sizes. For instance, the RCF of 1000 x g means that the centrifugal force is 1000 times stronger than the Earth's gravitational force. RCF is dependent on the speed of rotation in rpm and the distance of the particles from the center of rotation. The most common formula used for calculating RCF is:

$$RCF = 1.118 \times 10^{-5} \times r \times (rpm)^2$$

where $1.118 \times 10^{-5}$ is a constant; *r* is the radius, expressed in centimetres, between the axis of rotation and a point in the sample; and *rpm* is the speed in revolutions per minute.

Historically, many separations have been carried out at the speed of 3000 rpm; a rough guide to the 'g' force exerted at this speed is to multiply the centrifugation radius by a factor of 10, so a radius of 160 mm gives approximately 1600 x g. This is a rather arbitrary approach, since the RCF applied is linearly dependent on the radius, so a 10% larger radius means that a 10% higher RCF is applied at the same speed. Roughly, the above formula can be simplified to $RCF = 10 \times r$ , with an error of only 0.62%.

## Centrifugation in Biological Research

### Microcentrifuges

Microcentrifuges are specially designed table-top models with light, small-volume rotors capable of very fast acceleration up to approximately 17,000 rpm. They are lightweight devices which are primarily used for short-time centrifugation of samples up to around 0.2–2.0 mL. However, due to their small scale, they are readily transportable and, if necessary, can be operated in a cold room. They can be refrigerated or not. The microcentrifuge is normally used in research laboratories where small samples of biological molecules, cells, or nuclei are required to be subjected to high RCF for relatively short time intervals. Microcentrifuges designed for high-speed operation can reach up to 35,000 rpm, giving RCF up to 30000×g, and are called high-speed microcentrifuges.

### Low-Speed Centrifuges

Low-speed centrifuges are used to harvest chemical precipitates, intact cells (animal, plant and some microorganisms), nuclei, chloroplasts, large mitochondria and the larger plasma-membrane fragments. Density gradients for purifying cells are also run in these centrifuges. Swinging-bucket rotors tend to be used very widely because of the huge flexibility of sample size through the use of adaptors. These machines have maximum rotor speeds of less than 10 000 rpm and vary from small, bench-top to large, floor-standing centrifuges.

### High-Speed Centrifuges

High-speed centrifuges are typically used to harvest microorganisms, viruses, mitochondria, lysosomes, peroxisomes and intact tubular Golgi membranes. The majority of the simple pelleting tasks are carried out in fixed angle rotors. Some density-gradient work for purifying cells and organelles can be carried out in swinging-bucket rotors, or in the case of Percoll gradients in fixed-angle rotors. High-speed or superspeed centrifuges can handle larger sample volumes, from a few tens of millilitres to several litres. Additionally, larger centrifuges can also reach higher angular velocities (around 30,000 rpm). The rotors may come with different adapters to hold various sizes of test tubes, bottles, or microtiter plates.

### Ultracentrifugations

Ultracentrifugation makes use of high centrifugal force for studying properties of biological particles at exceptionally high speeds. Current ultracentrifuges can spin to as much as 150,000 rpm (equivalent to 1,000,000 x g). They are used to harvest all membrane vesicles derived from the plasma membrane, endoplasmic reticulum (ER) and Golgi membrane, endosomes, ribosomes, ribosomal subunits, plasmids, DNA, RNA and proteins in fixed-angle rotors. Compared to microcentrifuges or high-speed centrifuges, ultracentrifuges can isolate much smaller particles and, additionally, whilst microcentrifuges and supercentrifuges separate particles in batches (limited volumes of samples must be handled manually in test tubes or bottles), ultracentrifuges can separate molecules in batch or continuous flow systems.

Ultracentrifugation is employed for separation of macromolecules/ligand binding kinetic studies, separation of various lipoprotein fractions from plasma and deprotonisation of physiological fluids for amino acid analysis.

They are the most commonly used centrifuge for the density-gradient purification of all particles except cells, and, whilst swinging buckets have been traditionally used for this purpose, fixed-angle rotors and vertical rotors are also used, particularly for self-generated gradients and can improve the efficiency of separation greatly. There are two kinds of ultracentrifuges: the analytical and the preparative.

### Analytical Ultracentrifugation

Analytical ultracentrifugation (AUC) can be used for determination of the properties of macromolecules such as shape, mass, composition, and conformation. It is a commonly used biomolecular analysis technique used to evaluate sample purity, to characterize the assembly and disassembly mechanisms of biomolecular complexes, to determine subunit stoichiometries, to identify and characterize macromolecular conformational changes, and to calculate equilibrium constants and thermodynamic parameters for self-associating and hetero-associating systems. Analytical ultracentrifuges incorporate a scanning visible/ultraviolet light-based optical detection system for real-time monitoring of the sample's progress during a spin.

Samples are centrifuged with a high-density solution such as sucrose, caesium chloride, or iodixanol. The high-density solution may be at a uniform concentration throughout the test tube ("cushion") or a varying concentration ("gradient"). Molecular properties can be modeled through sedimentation velocity analysis or sedimentation equilibrium analysis. During the run, the particle or molecules will migrate through the test tube at different speeds depending on their physical properties and the properties of the solution, and eventually form a pellet at the bottom of the tube, or bands at various heights.

### Preparative Ultracentrifugation

Preparative ultracentrifuges are often used for separating particles according to their densities, isolating and/or harvesting denser particles for collection in the pellet, and clarifying suspensions containing particles. Sometimes researchers also use preparative ultracentrifuges if they need the flexibility to change the type of rotor in the instrument. Preparative ultracentrifuges can be equipped with a wide range of different rotor types, which can spin samples of different numbers, at different angles, and at different speeds.

### Fractionation Process

In biological research, cell fractionation typically includes the isolation of cellular components while retaining the individual roles of each component. Generally, the cell sample is stored in a suspension which is:

- **Buffered:** Neutral pH, preventing damage to the structure of proteins including enzymes (which could affect ionic bonds).
- **Isotonic (of Equal Water Potential):** This prevents water gain or loss by the organelles.
- **Cool:** Reducing the overall activity of enzyme released later in the procedure.

Centrifugation is the first step in most fractionations. Through low-speed centrifugation, cell debris may be removed, leaving a supernatant preserving the contents of the cell. Repeated centrifugation at progressively higher speeds will fractionate homogenates of cells into their components. In general, the smaller the subcellular component, the greater is the centrifugal force required to sediment it. The soluble fraction of any lysate can then be further separated into its constituents using a variety of methods.

### Differential Centrifugation

Differential centrifugation is the simplest method of fractionation by centrifugation, commonly used to separate organelles and membranes found in cells. Organelles generally differ from each other in density and in size, making the use of differential centrifugation, and centrifugation in general, possible. The organelles can then be identified by testing for indicators that are unique to the specific organelles. The most widely used application of this technique is to produce crude subcellular fractions from a tissue homogenate such as that from rat liver. Particles of different densities or sizes in a suspension are sedimented at different rates, with the larger and denser particles sedimenting faster. These sedimentation rates can be increased by using centrifugal force.

A suspension of cells is subjected to a series of increasing centrifugal force cycles to produce a series of pellets comprising cells with a declining sedimentation rate. Homogenate includes nuclei, mitochondria, lysosomes, peroxisomes, plasma membrane sheets and a broad range of vesicles derived from a number of intracellular membrane compartments and also from the plasma membrane, typically in a buffered medium.

### Density Gradient Centrifugation

Density gradient centrifugation is known to be one of the most efficient methods for separating suspended particles, and is used both as a separation technique and as a method for measuring the density of particles or molecules in a mixture.

It is used to separate particles on the basis of size, shape, and density by using a medium of graded densities. During a relatively short or slow centrifugation, the particles are separated by size, with larger particles sedimenting farther than smaller ones. Over a long or fast centrifugation, particles travel to locations in the gradient where the density of the medium is the same as that of the particle density; $(\rho p - \rho m) \rightarrow 0$. Therefore, a small, dense particle initially sediments less readily than a large, low density particle. The large particles reach their equilibrium density position early, while the small particles slowly migrate across the large particle zone and ultimately take up an equilibrium position deeper into the gradient.

A tube, after being centrifuged by this method, has particles in order of density based on height. The object or particle of interest will reside in the position within the tube corresponding to its density. Nevertheless, some non-ideal sedimentations are still possible when using this method. The first potential issue is the unwanted aggregation of particles, but this can occur in any centrifugation. The second possibility occurs when droplets of solution that contain particles sediment. This is more likely to occur when working with a solution that has a layer of suspension floating on a dense liquid, which in fact have little to no density gradient.

## Other Applications

A centrifuge can be used to isolate small quantities of solids retained in suspension from liquids, such as in the separation of chalk powder from water. In biological research, it can be used in the purification of mammalian cells, fractionation of subcellular organelles, fractionation of membrane vesicles, fractionation of macromolecules and macromolecular complexes, etc. Centrifugation is used in many different ways in the food industry. For example, in the dairy industry, it is typically used in the clarification and skimming of milk, extraction of cream, production and recovery of casein, cheese production, removing bacterial contaminants, etc. This processing technique is also used in the production of beverages, juices, coffee, tea, beer, wine, soy milk, oil and fat processing/recovery, cocoa butter, sugar production, etc. It is also used in the clarification and stabilization of wine. In forensic and research laboratories, it can be used in the separation of urine and blood components. It also aids in separation of proteins using purification techniques such as salting out, e.g., ammonium sulfate precipitation. Centrifugation is also an important technique in waste treatment, being one of the most common processes used for sludge dewatering. This process also plays a role in cyclonic separation, where particles are separated from an air-flow without the use of filters. In a cyclone collector, air moves in a helical path. Particles with high inertia are separated by the centrifugal force whilst smaller particles continue with the air-flow. Centrifuges have also been used to a small degree to isolate lighter-than-water compounds, such as oil. In such situations, the aqueous discharge is obtained at the opposite outlet from which solids with a specific gravity greater than one are the target substances for separation.

### 5.3.3 Fractionation

Fractionation is a separation process in which a certain quantity of a mixture (of gasses, solids, liquids, enzymes, or isotopes, or a suspension) is divided during a phase transition, into a number of smaller quantities (fractions) in which the composition varies according to a gradient. Fractions are collected based on differences in a specific property of the individual components. A common trait in fractionations is the need to find an optimum between the amount of fractions collected and the desired purity in each fraction. Fractionation makes it possible to isolate more than two components in a mixture in a single run. This property sets it apart from other separation techniques. Fractionation is widely employed in many branches of science and technology. Mixtures of liquids and gasses are separated by fractional distillation by difference in boiling point. Fractionation of components also takes place in column chromatography by a difference in affinity between stationary phase and the mobile phase. In fractional crystallization and fractional freezing, chemical substances are fractionated based on difference in solubility at a given temperature. In cell fractionation, cell components are separated by difference in mass.

#### Natural Samples

##### Bioassay-Guided Fractionation

A typical protocol to isolate a pure chemical agent from natural origin is step-by-step separation of extracted components based on differences in their bioassay-guided fractionation physicochemical properties, and assessing the biological activity, followed by next round of separation and assaying. Typically, such work is initiated after a given crude extract is deemed “active” in a particular in vitro assay.

##### Blood Fractionation

The process of blood fractionation involves separation of blood into its main components. Blood fractionation refers generally to the process of separation using a centrifuge (centrifugation), after which three major blood components can be visualized: plasma, buffy coat and erythrocytes (blood cells). These separated components can be analyzed and often further separated.

### Food

Fractionation is also used for culinary purposes, as coconut oil, palm oil, and palm kernel oil are fractionated to produce oils of different viscosities, that may be used for different purposes. These oils typically use fractional crystallization (separation by solubility at temperatures) for the separation process instead of distillation. Mango oil is an oil fraction obtained during the processing of mango butter. Milk can also be fractionated to recover the milk protein concentrate or the milk basic proteins fraction.

## 5.4 PROTEIN PURIFICATION TECHNIQUES

After extraction, various purification techniques are applied to isolate the target protein. These techniques often involve multiple steps to achieve the desired purity and quantity.

### 5.4.1 Chromatography

Chromatography is a versatile method for protein purification, with different forms tailored to separate proteins based on unique properties.

- **Affinity Chromatography**: Affinity chromatography uses a specific ligand attached to a column matrix to capture the target protein. For example, His-tagged proteins can bind to nickel columns, allowing for selective purification. Affinity chromatography is highly specific, often achieving significant purity in a single step.
- **Ion-Exchange Chromatography**: This technique separates proteins based on charge differences. Proteins bind to positively or negatively charged beads, depending on their isoelectric point (pI), and are eluted by changing the pH or salt concentration of the buffer. Ion-exchange chromatography is widely used for proteins with slight charge differences.
- **Size-Exclusion Chromatography (SEC)**: Also known as gel filtration chromatography, SEC separates proteins based on their size. Larger molecules elute faster than smaller ones, which enter the porous beads in the column. SEC is often used as a final polishing step to remove aggregates and achieve high purity.

### 5.4.2 Electrophoresis

While often used for analytical purposes, electrophoresis can also serve preparative functions for smaller-scale purifications.

- **SDS-PAGE**: Sodium dodecyl sulfate polyacrylamide gel electrophoresis

(SDS-PAGE) is commonly used to analyze protein purity. Although primarily analytical, it can also be used to purify proteins in small quantities by excising bands of interest.

- **Isoelectric Focusing (IEF):** IEF separates proteins based on their pI by applying an electric field across a pH gradient. Proteins migrate to the region where the pH matches their pI, allowing for purification based on charge.

### 5.4.3 Precipitation Methods

Protein precipitation is a valuable technique for concentrating proteins and removing contaminants.

- **Ammonium Sulfate Precipitation:** Ammonium sulfate is added to a protein solution to selectively precipitate proteins based on their solubility. This "salting out" process can concentrate proteins and remove contaminants.
- **Organic Solvent Precipitation:** Organic solvents like ethanol or acetone can also precipitate proteins. These methods are particularly useful for purifying proteins that are sensitive to salt.

## 5.5 QUALITY CONTROL AND ASSESSMENT OF PURIFIED PROTEINS

Once the target protein is purified, it is essential to assess its quality and ensure that it retains its functional properties. Several techniques are used to verify protein purity, identity, and functionality.

- **SDS-PAGE and Western Blotting:** SDS-PAGE provides a visual confirmation of protein purity by separating proteins based on size. Western blotting, which uses antibodies specific to the target protein, confirms its identity and helps detect any remaining impurities.
- **Mass Spectrometry:** Mass spectrometry can identify and verify the molecular weight of purified proteins. It is particularly useful for confirming the sequence of recombinant proteins and ensuring there are no unexpected modifications.
- **Enzyme Activity Assays:** For enzymatic proteins, activity assays provide a direct measure of function. By comparing activity before and after purification, researchers can verify that the protein remains functional and that it has not been denatured.
- **Structural Analysis:** For structural studies, techniques like circular dichroism (CD) spectroscopy or X-ray crystallography can assess protein folding and structure, confirming that the protein is in its native conformation.

## 5.6 APPLICATIONS OF PREPARATIVE PROTEIN BIOCHEMISTRY

The field of preparative protein biochemistry has wide-ranging applications across multiple scientific disciplines, enabling advances in medicine, biotechnology, and basic research.

- **Drug Development and Therapeutics**: Preparative protein biochemistry is instrumental in developing therapeutic proteins, such as monoclonal antibodies and enzyme replacement therapies. Purified proteins are tested for therapeutic efficacy and safety, enabling the production of drugs that target specific diseases.
- **Structural Biology**: Purified proteins are essential for structural studies, which reveal the three-dimensional arrangement of atoms within a molecule. Structural insights are critical for drug design, as they allow scientists to understand how proteins interact with drugs and other molecules.
- **Industrial Biotechnology**: In biotechnology, enzymes are used in industrial applications, such as biofuel production and food processing. Preparative protein biochemistry allows for the isolation and purification of enzymes in quantities sufficient for industrial use, providing efficient and sustainable solutions.
- **Basic Research**: Purified proteins are essential for studying fundamental cellular processes, including signal transduction, gene expression, and metabolic pathways. Preparative protein biochemistry enables scientists to explore these processes at a molecular level, enhancing our understanding of biology.

## 5.7 CONCLUSION

Preparative protein biochemistry is a foundational field that enables the isolation, purification, and functional analysis of proteins. By utilizing various techniques such as chromatography, electrophoresis, and precipitation, scientists can purify proteins in sufficient quantities for research, industrial, and therapeutic applications. As our understanding of protein function grows, so too does the importance of these preparative techniques in advancing scientific knowledge and developing new treatments. With ongoing innovations in purification methods and quality control, preparative protein biochemistry will continuc to play a vital role in biotechnology, drug development, and life sciences research.

# CHAPTER 6

# ELECTROPHORETIC TECHNIQUES

Electrophoresis is a fundamental method used in molecular biology, biochemistry, and genetics to separate and analyze molecules based on their charge and size. This powerful technique has revolutionized the study of proteins, nucleic acids, and other biomolecules, making it possible to study their properties, identify variations, and assess purity.

Electrophoretic techniques, including agarose gel electrophoresis, polyacrylamide gel electrophoresis, and capillary electrophoresis, among others, are crucial for research, diagnostics, and forensic applications.

## 6.1 PRINCIPLES OF ELECTROPHORESIS

Electrophoresis is based on the movement of charged molecules in an electric field. When an electric current is applied to a gel or other medium, molecules with positive or negative charges migrate towards the oppositely charged electrodes. The rate of migration depends on factors such as the molecule's charge, size, shape, and the properties of the medium. In electrophoretic separation, smaller molecules generally move faster than larger ones due to less frictional resistance. The choice of gel and buffer is important, as they influence both the separation resolution and the stability of the biomolecules.

- **Gel Matrices:** Gels, like agarose and polyacrylamide, are commonly used as matrices in electrophoresis. These gels create a mesh-like structure that influences the movement of molecules based on their size. Agarose gels are typically used for nucleic acids, while polyacrylamide gels are often used for proteins due to their smaller pore size, which allows for higher resolution.
- **Buffers:** Buffers maintain a stable pH and provide ions that conduct electricity, ensuring consistent migration of molecules. Different buffers are used for different applications, and choosing the appropriate buffer is crucial for accurate and reproducible results.

## 6.2 TYPES OF ELECTROPHORETIC TECHNIQUES

Various electrophoretic techniques are used to analyze proteins, nucleic acids, and other biomolecules, each with unique applications and benefits.

### 6.2.1 Agarose Gel Electrophoresis

Agarose gel electrophoresis is a method of gel electrophoresis used in biochemistry, molecular biology, genetics, and clinical chemistry to separate a mixed population of macromolecules such as DNA or proteins in a matrix of agarose, one of the two main components of agar. The proteins may be separated by charge and/or size (isoelectric focusing agarose electrophoresis is essentially size independent), and the DNA and RNA fragments by length. Biomolecules are separated by applying an electric field to move the charged molecules through an agarose matrix, and the biomolecules are separated by size in the agarose gel matrix.

Agarose gel is easy to cast, has relatively fewer charged groups, and is particularly suitable for separating DNA of size range most often encountered in laboratories, which accounts for the popularity of its use. The separated DNA may be viewed with stain, most commonly under UV light, and the DNA fragments can be extracted from the gel with relative ease. Most agarose gels used are between 0.7–2% dissolved in a suitable electrophoresis buffer.

#### Properties of Agarose Gel

Agarose gel is a three-dimensional matrix formed of helical agarose molecules in supercoiled bundles that are aggregated into three-dimensional structures with channels and pores through which biomolecules can pass. The 3-D structure is held together with hydrogen bonds and can therefore be disrupted by heating back to a liquid state.

The melting temperature is different from the gelling temperature, depending on the sources, agarose gel has a gelling temperature of 35–42 °C and a melting temperature of 85–95 °C. Low-melting and low-gelling agaroses made through chemical modifications are also available.

Agarose gel has large pore size and good gel strength, making it suitable as an anticonvection medium for the electrophoresis of DNA and large protein molecules. The pore size of a 1% gel has been estimated from 100 nm to 200–500 nm, and its gel strength allows gels as dilute as 0.15% to form a slab for gel electrophoresis. Low-concentration gels (0.1–0.2%) however are fragile and therefore hard to handle. Agarose gel has lower resolving power than polyacrylamide gel for DNA but has a greater range of separation, and is therefore used for DNA fragments of usually 50–20,000 bp in size.

The limit of resolution for standard agarose gel electrophoresis is around 750 kb, but resolution of over 6 Mb is possible with pulsed field gel electrophoresis (PFGE). It can also be used to separate large proteins, and it is the preferred matrix for the gel electrophoresis of particles with effective radii larger than 5–10 nm. A 0.9% agarose gel has pores large enough for the entry of bacteriophage T4.

The agarose polymer contains charged groups, in particular pyruvate and sulphate. These negatively charged groups create a flow of water in the opposite direction to the movement of DNA in a process called electroendosmosis (EEO), and can therefore retard the movement of DNA and cause blurring of bands. Higher concentration gels would have higher electroendosmotic flow. Low EEO agarose is therefore generally preferred for use in agarose gel electrophoresis of nucleic acids, but high EEO agarose may be used for other purposes.

The lower sulphate content of low EEO agarose, particularly low-melting point (LMP) agarose, is also beneficial in cases where the DNA extracted from gel is to be used for further manipulation as the presence of contaminating sulphates may affect some subsequent procedures, such as ligation and PCR. Zero EEO agaroses however are undesirable for some applications as they may be made by adding positively charged groups and such groups can affect subsequent enzyme reactions. Electroendosmosis is a reason agarose is used in preference to agar as the agaropectin component in agar contains a significant amount of negatively charged sulphate and carboxyl groups.

The removal of agaropectin in agarose substantially reduces the EEO, as well as reducing the non-specific adsorption of biomolecules to the gel matrix. However, for some applications such as the electrophoresis of serum proteins, a high EEO may be desirable, and agaropectin may be added in the gel used.

## Migration of Nucleic Acids in Agarose Gel

A number of factors can affect the migration of nucleic acids: the dimension of the gel pores (gel concentration), size of DNA being electrophoresed, the voltage used, the ionic strength of the buffer, and the concentration of intercalating dye such as ethidium bromide if used during electrophoresis.

Smaller molecules travel faster than larger molecules in gel, and double-stranded DNA moves at a rate that is inversely proportional to the logarithm of the number of base pairs. This relationship however breaks down with very large DNA fragments, and separation of very large DNA fragments requires the use of pulsed field gel electrophoresis (PFGE), which applies alternating current from different directions and the large DNA fragments are separated as they reorient themselves with the changing field.

For standard agarose gel electrophoresis, larger molecules are resolved better using a low concentration gel while smaller molecules separate better at high concentration gel. Higher concentration gels, however, require longer run times (sometimes days).

The movement of the DNA may be affected by the conformation of the DNA molecule, for example, supercoiled DNA usually moves faster than relaxed DNA because it is tightly coiled and hence more compact. In a normal plasmid DNA preparation, multiple forms of DNA may be present. Gel electrophoresis of the plasmids would normally show the negatively supercoiled form as the main band, while nicked DNA (open circular form) and the relaxed closed circular form appears as minor bands. The rate at which the various forms move however can change using different electrophoresis conditions, and the mobility of larger circular DNA may be more strongly affected than linear DNA by the pore size of the gel.

Ethidium bromide which intercalates into circular DNA can change the charge, length, as well as the superhelicity of the DNA molecule, therefore its presence in gel during electrophoresis can affect its movement. For example, the positive charge of ethidium bromide can reduce the DNA movement by 15%. Agarose gel electrophoresis can be used to resolve circular DNA with different supercoiling topology.

DNA damage due to increased cross-linking will also reduce electrophoretic DNA migration in a dose-dependent way. The rate of migration of the DNA is proportional to the voltage applied, i.e. the higher the voltage, the faster the DNA moves. The resolution of large DNA fragments however is lower at high voltage. The mobility of DNA may also change in an unsteady field – in a field that is periodically reversed, the mobility of DNA of a particular size may drop significantly at a particular cycling frequency. This phenomenon can result in band inversion in field inversion gel electrophoresis (FIGE), whereby larger DNA fragments move faster than smaller ones.

The negative charge of its phosphate backbone moves the DNA towards the positively charged anode during electrophoresis. However, the migration of DNA molecules in solution, in the absence of a gel matrix, is independent of molecular weight during electrophoresis. The gel matrix is therefore responsible for the separation of DNA by size during electrophoresis, and a number of models exist to explain the mechanism of separation of biomolecules in gel matrix. A widely accepted one is the Ogston model which treats the polymer matrix as a sieve. A globular protein or a random coil DNA moves through the interconnected pores, and the movement of larger molecules is more likely to be impeded and slowed down by collisions with the gel matrix, and the molecules of different sizes can therefore be separated in this sieving process.

The Ogston model however breaks down for large molecules whereby the pores are significantly smaller than size of the molecule. For DNA molecules of size greater than 1 kb, a reptation model (or its variants) is most commonly used. This model assumes that the DNA can crawl in a "snake-like" fashion (hence "reptation") through the pores as an elongated molecule. A biased reptation model applies at higher electric field strength, whereby the leading end of the molecule become strongly biased in the forward direction and pulls the rest of the molecule along. Real-time fluorescence microscopy of stained molecules, however, showed more subtle dynamics during electrophoresis, with the DNA showing considerable elasticity as it alternately stretching in the direction of the applied field and then contracting into a ball, or becoming hooked into a U-shape when it gets caught on the polymer fibres.

## General Procedure

The gel is prepared by dissolving the agarose powder in an appropriate buffer, such as TAE or TBE, to be used in electrophoresis. The agarose is dispersed in the buffer before heating it to near-boiling point, but avoid boiling. The melted agarose is allowed to cool sufficiently before pouring the solution into a cast as the cast may warp or crack if the agarose solution is too hot. A comb is placed in the cast to create wells for loading sample, and the gel should be completely set before use.

The concentration of gel affects the resolution of DNA separation. The agarose gel is composed of microscopic pores through which the molecules travel, and there is an inverse relationship between the pore size of the agarose gel and the concentration - pore size decreases as the density of agarose fibers increases. High gel concentration improves separation of smaller DNA molecules, while lowering gel concentration permits large DNA molecules to be separated. The process allows fragments ranging from 50 base pairs to several mega bases to be separated depending on the gel concentration used. The concentration is measured in weight of agarose over volume of buffer used (g/ml). For a standard agarose gel electrophoresis, a 0.8% gel gives good separation or resolution of large 5–10kb DNA fragments, while 2% gel gives good resolution for small 0.2–1kb fragments. 1% gels is often used for a standard electrophoresis. High percentage gels are often brittle and may not set evenly, while low percentage gels (0.1-0.2%) are fragile and not easy to handle. Low-melting-point (LMP) agarose gels are also more fragile than normal agarose gel. Low-melting point agarose may be used on its own or simultaneously with standard agarose for the separation and isolation of DNA. PFGE and FIGE are often done with high percentage agarose gels.

Once the gel has set, the comb is removed, leaving wells where DNA samples can be loaded. Loading buffer is mixed with the DNA sample before the mixture is loaded into the wells.

The loading buffer contains a dense compound, which may be glycerol, sucrose, or Ficoll, that raises the density of the sample so that the DNA sample may sink to the bottom of the well. If the DNA sample contains residual ethanol after its preparation, it may float out of the well. The loading buffer also includes colored dyes such as xylene cyanol and bromophenol blue used to monitor the progress of the electrophoresis. The DNA samples are loaded using a pipette.

Agarose gel electrophoresis is most commonly done horizontally in a subaqueous mode whereby the slab gel is completely submerged in buffer during electrophoresis. It is also possible, but less common, to perform the electrophoresis vertically, as well as horizontally with the gel raised on agarose legs using an appropriate apparatus. The buffer used in the gel is the same as the running buffer in the electrophoresis tank, which is why electrophoresis in the subaqueous mode is possible with agarose gel.

For optimal resolution of DNA greater than 2 kb in size in standard gel electrophoresis, 5 to 8 V/cm is recommended (the distance in cm refers to the distance between electrodes, therefore this recommended voltage would be 5 to 8 multiplied by the distance between the electrodes in cm). Voltage may also be limited by the fact that it heats the gel and may cause the gel to melt if it is run at high voltage for a prolonged period, especially if the gel used is LMP agarose gel. Too high a voltage may also reduce resolution, as well as causing band streaking for large DNA molecules. Too low a voltage may lead to broadening of band for small DNA fragments due to dispersion and diffusion.

Since DNA is not visible in natural light, the progress of the electrophoresis is monitored using colored dyes. Xylene cyanol (light blue color) comigrates large DNA fragments, while Bromophenol blue (dark blue) comigrates with the smaller fragments. Less commonly used dyes include Cresol Red and Orange G which migrate ahead of bromophenol blue.

A DNA marker is also run together for the estimation of the molecular weight of the DNA fragments. Note however that the size of a circular DNA like plasmids cannot be accurately gauged using standard markers unless it has been linearized by restriction digest, alternatively a supercoiled DNA marker may be used.

DNA as well as RNA are normally visualized by staining with ethidium bromide, which intercalates into the major grooves of the DNA and fluoresces under UV light. The intercalation depends on the concentration of DNA and thus, a band with high intensity will indicate a higher amount of DNA compared to a band of less intensity.

The ethidium bromide may be added to the agarose solution before it gels, or the DNA gel may be stained later after electrophoresis. Destaining of the gel is not necessary but may produce better images. Other methods of staining are available; examples are MIDORI Green, SYBR Green, GelRed, methylene blue, brilliant cresyl blue, Nile blue sulphate, and crystal violet. SYBR Green, GelRed and other similar commercial products are sold as safer alternatives to ethidium bromide as it has been shown to be mutagenic in Ames test, although the carcinogenicity of ethidium bromide has not actually been established. SYBR Green requires the use of a blue-light transilluminator. DNA stained with crystal violet can be viewed under natural light without the use of a UV transilluminator which is an advantage, however it may not produce a strong band.

When stained with ethidium bromide, the gel is viewed with an ultraviolet (UV) transilluminator. The UV light excites the electrons within the aromatic ring of ethidium bromide, and once they return to the ground state, light is released, making the DNA and ethidium bromide complex fluoresce. Standard transilluminators use wavelengths of 302/312-nm (UV-B), however exposure of DNA to UV radiation for as little as 45 seconds can produce damage to DNA and affect subsequent procedures, for example reducing the efficiency of transformation, *in vitro* transcription, and PCR. Exposure of DNA to UV radiation therefore should be limited. Using a higher wavelength of 365 nm (UV-A range) causes less damage to the DNA but also produces much weaker fluorescence with ethidium bromide. Where multiple wavelengths can be selected in the transilluminator, shorter wavelength can be used to capture images, while longer wavelength should be used if it is necessary to work on the gel for any extended period of time.

The transilluminator apparatus may also contain image capture devices, such as a digital or polaroid camera, that allow an image of the gel to be taken or printed. For gel electrophoresis of protein, the bands may be visualised with Coomassie or silver stains. The separated DNA bands are often used for further procedures, and a DNA band may be cut out of the gel as a slice, dissolved and purified. Contaminants however may affect some downstream procedures such as PCR, and low melting point agarose may be preferred in some cases as it contains fewer of the sulphates that can affect some enzymatic reactions. The gels may also be used for blotting techniques.

## Buffers

In general, the ideal buffer should have good conductivity, produce less heat and have a long life. There are a number of buffers used for agarose electrophoresis; common ones for nucleic acids include Tris/Acetate/EDTA (TAE) and Tris/Borate/EDTA (TBE). The buffers used contain EDTA to inactivate many nucleases which require divalent cation for their function. The borate in TBE buffer can be problematic as borate can polymerize, and/or interact with cis diols such as those found in RNA. TAE has the lowest buffering capacity, but it provides the best resolution for larger DNA. This means a lower voltage and more time, but a better product.

Many other buffers have been proposed, e.g. lithium borate (LB), iso electric histidine, pK matched goods buffers, etc.; in most cases the purported rationale is lower current (less heat) and or matched ion mobilities, which leads to longer buffer life. Tris-phosphate buffer has high buffering capacity but cannot be used if DNA extracted is to be used in phosphate sensitive reaction. LB is relatively new and is ineffective in resolving fragments larger than 5 kbp; However, with its low conductivity, a much higher voltage could be used (up to 35 V/cm), which means a shorter analysis time for routine electrophoresis. As low as one base pair size difference could be resolved in 3% agarose gel with an extremely low conductivity medium (1 mM lithium borate).

Other buffering system may be used in specific applications, for example, barbituric acid-sodium barbiturate or Tris-barbiturate buffers may be used for in agarose gel electrophoresis of proteins, for example in the detection of abnormal distribution of proteins.

### Applications

- Estimation of the size of DNA molecules following digestion with restriction enzymes, e.g., in restriction mapping of cloned DNA.
- Estimation of the DNA concentration by comparing the intensity of the nucleic acid band with the corresponding band of the size marker.
- Analysis of products of a polymerase chain reaction (PCR), e.g., in molecular genetic diagnosis or genetic fingerprinting.
- Separation of DNA fragments for extraction and purification.
- Separation of restricted genomic DNA prior to Southern transfer, or of RNA prior to Northern transfer.
- Separation of proteins, for example, screening of protein abnormalities in clinical chemistry.

Agarose gels are easily cast and handled compared to other matrices and nucleic acids are not chemically altered during electrophoresis. Samples are also easily recovered. After the experiment is finished, the resulting gel can be stored in a plastic bag in a refrigerator.

Electrophoresis is performed in buffer solutions to reduce pH changes due to the electric field, which is important because the charge of DNA and RNA depends on pH, but running for too long can exhaust the buffering capacity of the solution. Further, different preparations of genetic material may not migrate consistently with each other, for morphological or other reasons.

## 6.2.2 Polyacrylamide Gel Electrophoresis (PAGE)

Polyacrylamide gel electrophoresis (PAGE) is a technique widely used in biochemistry, forensic chemistry, genetics, molecular biology and biotechnology to separate biological macromolecules, usually proteins or nucleic acids, according to their electrophoretic mobility. Electrophoretic mobility is a function of the length, conformation, and charge of the molecule. Polyacrylamide gel electrophoresis is a powerful tool used to analyze RNA samples. When polyacrylamide gel is denatured after electrophoresis, it provides information on the sample composition of the RNA species.

Hydration of acrylonitrile results in formation of acrylamide molecules ($C_3H_5NO$) by nitrile hydratase. Acrylamide monomer is in a powder state before addition of water. Acrylamide is toxic to the human nervous system, therefore all safety measures must be followed when working with it. Acrylamide is soluble in water and upon addition of free-radical initiators it polymerizes resulting in formation of polyacrylamide. It is useful to make polyacrylamide gel via acrylamide hydration because pore size can be regulated. Increased concentrations of acrylamide result in decreased pore size after polymerization. Polyacrylamide gel with small pores helps to examine smaller molecules better since the small molecules can enter the pores and travel through the gel while large molecules get trapped at the pore openings.

As with all forms of gel electrophoresis, molecules may be run in their native state, preserving the molecules' higher-order structure. This method is called native-PAGE. Alternatively, a chemical denaturant may be added to remove this structure and turn the molecule into an unstructured molecule whose mobility depends only on its length (because the protein-SDS complexes all have a similar mass-to-charge ratio).

This procedure is called SDS-PAGE. Sodium dodecyl sulfate polyacrylamide gel electrophoresis (SDS-PAGE) is a method of separating molecules based on the difference of their molecular weight. At the pH at which gel electrophoresis is carried out the SDS molecules are negatively charged and bind to proteins in a set ratio, approximately one molecule of SDS for every 2 amino acids. In this way, the detergent provides all proteins with a uniform charge-to-mass ratio. By binding to the proteins the detergent destroys their secondary, tertiary and/or quaternary structure denaturing them and turning them into negatively charged linear polypeptide chains. When subjected to an electric field in PAGE, the negatively charged polypeptide chains travel toward the anode with different mobility. Their mobility, or the distance traveled by molecules, is inversely proportional to the logarithm of their molecular weight. By comparing the relative ratio of the distance traveled by each protein to the length of the gel (Rf) one can make conclusions about the relative molecular weight of the proteins, where the length of the gel is determined by the distance traveled by a small molecule like a tracking dye.

For nucleic acids, urea is the most commonly used denaturant. For proteins, sodium dodecyl sulfate (SDS) is an anionic detergent applied to protein samples to coat proteins in order to impart two negative charges (from every SDS molecule) to every two amino acids of the denatured protein. 2-Mercaptoethanol may also be used to disrupt the disulfide bonds found between the protein complexes, which helps further denature the protein.

In most proteins, the binding of SDS to the polypeptide chains impart an even distribution of charge per unit mass, thereby resulting in a fractionation by approximate size during electrophoresis. Proteins that have a greater hydrophobic content – for instance, many membrane proteins, and those that interact with surfactants in their native environment – are intrinsically harder to treat accurately using this method, due to the greater variability in the ratio of bound SDS.

Procedurally, using both Native and SDS-PAGE together can be used to purify and to separate the various subunits of the protein. Native-PAGE keeps the oligomeric form intact and will show a band on the gel that is representative of the level of activity.

SDS-PAGE will denature and separate the oligomeric form into its monomers, showing bands that are representative of their molecular weights. These bands can be used to identify and assess the purity of the protein.

## Procedure

### Sample Preparation

Samples may be any material containing proteins or nucleic acids. These may be biologically derived, for example from prokaryotic or eukaryotic cells, tissues, viruses, environmental samples, or purified proteins.

In the case of solid tissues or cells, these are often first broken down mechanically using a blender (for larger sample volumes), using a homogenizer (smaller volumes), by sonicator or by using cycling of high pressure, and a combination of biochemical and mechanical techniques – including various types of filtration and centrifugation – may be used to separate different cell compartments and organelles prior to electrophoresis. Synthetic biomolecules such as oligonucleotides may also be used as analytes.

The sample to analyze is optionally mixed with a chemical denaturant if so desired, usually SDS for proteins or urea for nucleic acids. SDS is an anionic detergent that denatures secondary and non–disulfide–linked tertiary structures, and additionally applies a negative charge to each protein in proportion to its mass. Urea breaks the hydrogen bonds between the base pairs of the nucleic acid, causing the constituent strands to separate. Heating the samples to at least 60 °C further promotes denaturation.

In addition to SDS, proteins may optionally be briefly heated to near boiling in the presence of a reducing agent, such as dithiothreitol (DTT) or 2-mercaptoethanol (beta-mercaptoethanol/BME), which further denatures the proteins by reducing disulfide linkages, thus overcoming some forms of tertiary protein folding, and breaking up quaternary protein structure (oligomeric subunits). This is known as reducing SDS-PAGE.

A tracking dye may be added to the solution. This typically has a higher electrophoretic mobility than the analytes to allow the experimenter to track the progress of the solution through the gel during the electrophoretic run.

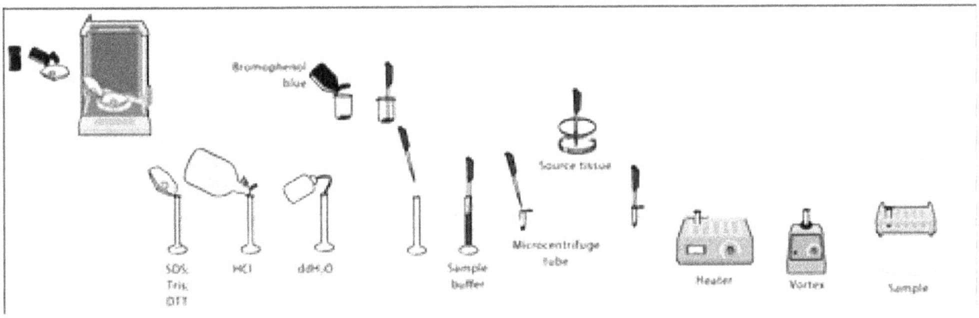


### Preparing Acrylamide Gels

The gels typically consist of acrylamide, bisacrylamide, the optional denaturant (SDS or urea), and a buffer with an adjusted pH. The solution may be degassed under a vacuum to prevent the formation of air bubbles during polymerization. Alternatively, butanol may be added to the resolving gel (for proteins) after it is poured, as butanol removes bubbles and makes the surface smooth.

A source of free radicals and a stabilizer, such as ammonium persulfate and TEMED are added to initiate polymerization. The polymerization reaction creates a gel because of the added bisacrylamide, which can form cross-links between two acrylamide molecules.

The ratio of bisacrylamide to acrylamide can be varied for special purposes, but is generally about 1 part in 35. The acrylamide concentration of the gel can also be varied, generally in the range from 5% to 25%. Lower percentage gels are better for resolving very high molecular weight molecules, while much higher percentages of acrylamide are needed to resolve smaller proteins.

The average pore diameter of polyacrylamide gels is determined by the total concentration of acrylamides (% T with T = Total concentration of acrylamide and bisacrylamide) and the concentration of the cross-linker bisacrylamide (%C with C = bisacrylamide concentration).

The pore size is reduced reciprocally to the %T. Concerning %C, a concentration of 5% produces the smallest pores, since the influence of bisacrylamide on the pore size has a parabola-shape with a vertex at 5%.

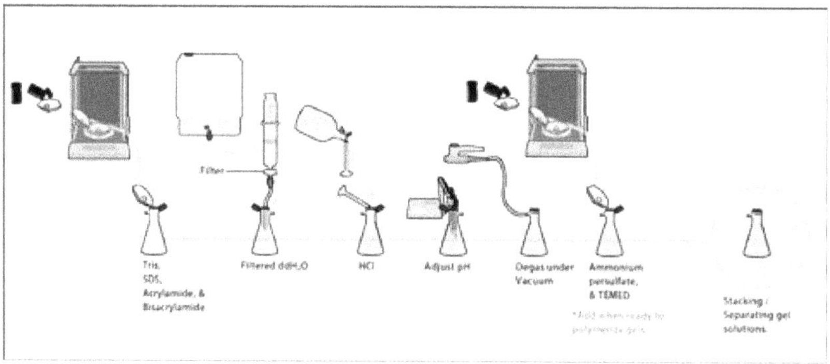


Gels are usually polymerized between two glass plates in a gel caster, with a comb inserted at the top to create the sample wells. After the gel is polymerized the comb can be removed and the gel is ready for electrophoresis.

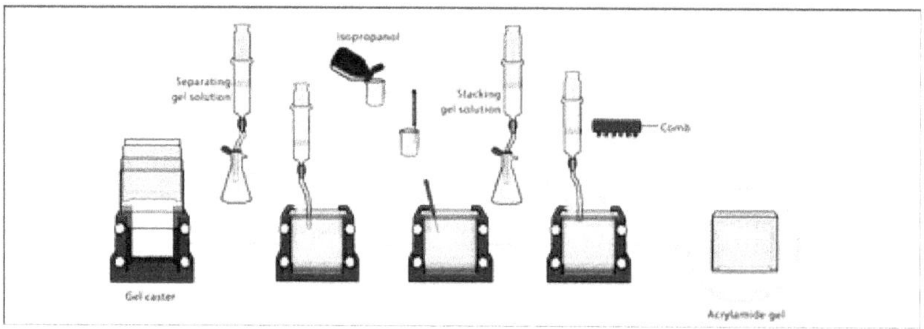


## Electrophoresis

Various buffer systems are used in PAGE depending on the nature of the sample and the experimental objective. The buffers used at the anode and cathode may be the same or different.

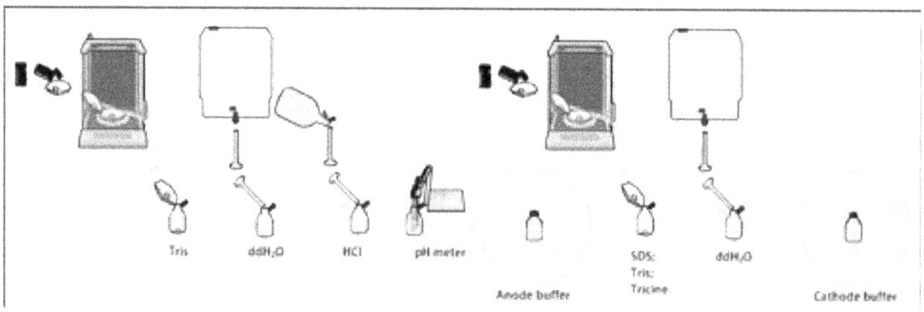


An electric field is applied across the gel, causing the negatively charged proteins or nucleic acids to migrate across the gel away from the negative electrode (which is the cathode being that this is an electrolytic rather than galvanic cell) and towards the positive electrode (the anode).

Depending on their size, each biomolecule moves differently through the gel matrix: small molecules more easily fit through the pores in the gel, while larger ones have more difficulty. The gel is run usually for a few hours, though this depends on the voltage applied across the gel; migration occurs more quickly at higher voltages, but these results are typically less accurate than at those at lower voltages.

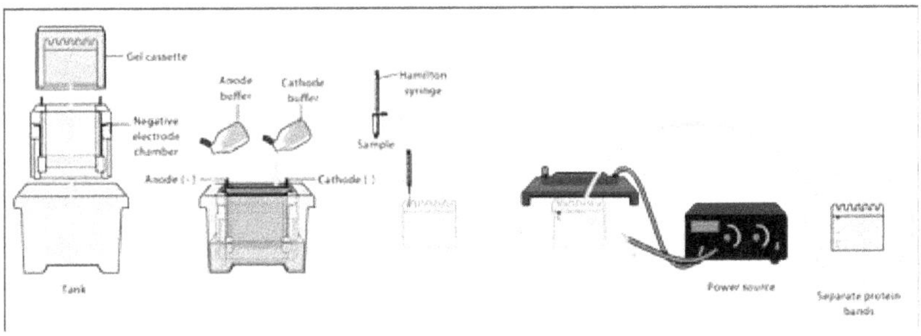


After the set amount of time, the biomolecules have migrated different distances based on their size. Smaller biomolecules travel farther down the gel, while larger ones remain closer to the point of origin. Biomolecules may therefore be separated roughly according to size, which depends mainly on molecular weight under denaturing conditions, but also depends on higher-order conformation under native conditions. The gel mobility is defined as the rate of migration traveled with a voltage gradient of 1V/cm and has units of $cm^2$/sec/V. For analytical purposes, the relative mobility of biomolecules, $R_f$, the ratio of the distance the molecule traveled on the gel to the total travel distance of a tracking dye is plotted versus the molecular weight of the molecule (or sometimes the log of MW, or rather the $M_r$, molecular radius). Such typically linear plots represent the standard markers or calibration curves that are widely used for the quantitative estimation of a variety of biomolecular sizes.

Certain glycoproteins, however, behave anomalously on SDS gels. Additionally, the analysis of larger proteins ranging from 250,000 to 600,000 Da is also reported to be problematic due to the fact that such polypeptides move improperly in the normally used gel systems.

## Further Processing

Following electrophoresis, the gel may be stained (for proteins, most commonly with Coomassie brilliant blue R-250 or autoradiography; for nucleic acids, ethidium bromide; or for either, silver stain), allowing visualization of the separated proteins, or processed further (e.g. Western blot). After staining, different species biomolecules appear as distinct bands within the gel. It is common to run molecular weight size markers of known molecular weight in a separate lane in the gel to calibrate the gel and determine the approximate molecular mass of unknown biomolecules by comparing the distance traveled relative to the marker.

For proteins, SDS-PAGE is usually the first choice as an assay of purity due to its reliability and ease. The presence of SDS and the denaturing step make proteins separate, approximately based on size, but aberrant migration of some proteins may occur. Different proteins may also stain differently, which interferes with quantification by staining. PAGE may also be used as a preparative technique for the purification of proteins. For example, preparative native PAGE is a method for separating native metalloproteins in complex biological matrices.

## Chemical Ingredients and their Roles

*Polyacrylamide gel (PAG)* had been known as a potential embedding medium for sectioning tissues as early as 1964, and two independent groups employed PAG in electrophoresis in 1959. It possesses several electrophoretically desirable features that make it a versatile medium. It is a synthetic, thermo-stable, transparent, strong, chemically relatively inert gel, and can be prepared with a wide range of average pore sizes.

The pore size of a gel and the reproducibility in gel pore size are determined by three factors, the total amount of acrylamide present (%T) (T = Total concentration of acrylamide and bisacrylamide monomer), the amount of cross-linker (%C) (C = bisacrylamide concentration), and the time of polymerization of acrylamide (cf. QPNC-PAGE). Pore size decreases with increasing %T; with cross-linking, 5%C gives the smallest pore size. Any increase or decrease in %C from 5% increases the pore size, as pore size with respect to %C is a parabolic function with vertex as 5%C. This appears to be because of non-homogeneous bundling of polymer strands within the gel. This gel material can also withstand high voltage gradients, is amenable to various staining and destaining procedures, and can be digested to extract separated fractions or dried for autoradiography and permanent recording.

### Components

Polyacrylamide gels are composed of a stacking gel and separating gel. Stacking gels have a higher porosity relative to the separating gel, and allow for proteins to migrate in a concentrated area. Additionally, stacking gels usually have a pH of 6.8, since the neutral glycine molecules allow for faster protein mobility. Separating gels have a pH of 8.8, where the anionic glycine slows down the mobility of proteins. Separating gels allow for the separation of proteins and have a relatively lower porosity. Here, the proteins are separated based on size (in SDS-PAGE) and size/charge (Native PAGE).

Chemical buffer stabilizes the pH value to the desired value within the gel itself and in the electrophoresis buffer. The choice of buffer also affects the electrophoretic mobility of the buffer counterions and thereby the resolution of the gel. The buffer should also be unreactive and not modify or react with most proteins. Different buffers may be used as cathode and anode buffers, respectively, depending on the application. Multiple pH values may be used within a single gel, for example in DISC electrophoresis. Common buffers in PAGE include Tris, Bis-Tris, or imidazole.

Counterion balance the intrinsic charge of the buffer ion and also affect the electric field strength during electrophoresis. Highly charged and mobile ions are often avoided in SDS-PAGE cathode buffers, but may be included in the gel itself, where it migrates ahead of the protein. In applications such as DISC SDS-PAGE the pH values within the gel may vary to change the average charge of the counterions during the run to improve resolution. Popular counterions are glycine and tricine. Glycine has been used as the source of trailing ion or slow ion because its pKa is 9.69 and mobility of glycinate are such that the effective mobility can be set at a value below that of the slowest known proteins of net negative charge in the pH range. The minimum pH of this range is approximately 8.0.

Acrylamide ($C_3H_5NO$; mW: 71.08) when dissolved in water, slow, spontaneous autopolymerization of acrylamide takes place, joining molecules together by head on tail fashion to form long single-chain polymers. The presence of a free radical-generating system greatly accelerates polymerization. This kind of reaction is known as vinyl addition polymerisation. A solution of these polymer chains becomes viscous but does not form a gel, because the chains simply slide over one another. Gel formation requires linking various chains together. Acrylamide is carcinogenic, a neurotoxin, and a reproductive toxin. It is also essential to store acrylamide in a cool dark and dry place to reduce autopolymerisation and hydrolysis.

Bisacrylamide (*N,N′*-Methylenebisacrylamide) ($C_7H_{10}N_2O_2$; mW: 154.17) is the most frequently used cross linking agent for polyacrylamide gels. Chemically it can be thought of as two acrylamide molecules coupled head to head at their non-reactive ends. Bisacrylamide can crosslink two polyacrylamide chains to one another, thereby resulting in a gel.

Sodium dodecyl sulfate (SDS) ($C_{12}H_{25}NaO_4S$; mW: 288.38) (only used in denaturing protein gels) is a strong detergent agent used to denature native proteins to individual polypeptides. This denaturation, which is referred to as reconstructive denaturation, is not accomplished by the total linearization of the protein, but instead, through a conformational change to a combination of random coil and $\alpha$ helix secondary structures. When a protein mixture is heated to 100°C in presence of SDS, the detergent wraps around the polypeptide backbone. It binds to polypeptides in a constant weight ratio of 1.4 g SDS/g of polypeptide. In this process, the intrinsic charges of polypeptides become negligible when compared to the negative charges contributed by SDS.

Thus polypeptides after treatment become rod-like structures possessing a uniform charge density, that is same net negative charge per unit weight. The electrophoretic mobilities of these proteins is a linear function of the logarithms of their molecular weights. Without SDS, different proteins with similar molecular weights would migrate differently due to differences in mass-charge ratio, as each protein has an isoelectric point and molecular weight particular to its primary structure. This is known as native PAGE. Adding SDS solves this problem, as it binds to and unfolds the protein, giving a near uniform negative charge along the length of the polypeptide.

Urea ($CO(NH_2)_2$; mW: 60.06) is a chaotropic agent that increases the entropy of the system by interfering with intramolecular interactions mediated by non-covalent forces such as hydrogen bonds and van der Waals forces. Macromolecular structure is dependent on the net effect of these forces, therefore it follows that an increase in chaotropic solutes denatures macromolecules,

Ammonium persulfate (APS) ($N_2H_8S_2O_8$; mW: 228.2) is a source of free radicals and is often used as an initiator for gel formation. An alternative source of free radicals is riboflavin, which generated free radicals in a photochemical reaction.

TEMED (*N, N, N′, N′*-tetramethylethylenediamine) ($C_6H_{16}N_2$; mW: 116.21) stabilizes free radicals and improves polymerization. The rate of polymerisation and the properties of the resulting gel depend on the concentrations of free radicals. Increasing the amount of free radicals results in a decrease in the average polymer chain length, an increase in gel turbidity and a decrease in gel elasticity. Decreasing the amount shows the reverse effect. The lowest catalytic concentrations that allow polymerisation in a reasonable period of time should be used. APS and TEMED are typically used at approximately equimolar concentrations in the range of 1 to 10 mM.

### Chemicals for Processing and Visualization

Tracking dye; as proteins and nucleic acids are mostly colorless, their progress through the gel during electrophoresis cannot be easily followed. Anionic dyes of a known electrophoretic mobility are therefore usually included in the PAGE sample buffer. A very common tracking dye is Bromophenol blue (BPB, 3',3",5',5" tetrabromophenolsulfonphthalein).

This dye is coloured at alkali and neutral pH and is a small negatively charged molecule that moves towards the anode. Being a highly mobile molecule it moves ahead of most proteins. As it reaches the anodic end of the electrophoresis medium electrophoresis is stopped. It can weakly bind to some proteins and impart a blue colour. Other common tracking dyes are xylene cyanol, which has lower mobility, and Orange G, which has a higher mobility.

Loading aids; most PAGE systems are loaded from the top into wells within the gel. To ensure that the sample sinks to the bottom of the gel, sample buffer is supplemented with additives that increase the density of the sample. These additives should be non-ionic and non-reactive towards proteins to avoid interfering with electrophoresis. Common additives are glycerol and sucrose.

Coomassie brilliant blue R-250 (CBB)($C_{45}H_{44}N_3NaO_7S_2$; mW: 825.97) is the most popular protein stain. It is an anionic dye, which non-specifically binds to proteins. The structure of CBB is predominantly non-polar, and it is usually used in methanolic solution acidified with acetic acid. Proteins in the gel are fixed by acetic acid and simultaneously stained. The excess dye incorporated into the gel can be removed by destaining with the same solution without the dye. The proteins are detected as blue bands on a clear background. As SDS is also anionic, it may interfere with staining process. Therefore, large volume of staining solution is recommended, at least ten times the volume of the gel.

Ethidium bromide (EtBr) is a popular nucleic acid stain. EtBr allows one to easily visualize DNA or RNA on a gel as EtBr fluoresces an orange color under UV light. Ethidium bromide binds nucleic acid chains through the process of Intercalation. While Ethidium bromide is a popular stain it is important to exercise caution when using EtBr as it is a known carcinogen. Because of this fact, many researchers opt to use stains such as SYBR Green and SYBR Safe which are safer alternatives to EtBr. EtBr is used by simply adding it to the gel mixture. Once the gel has run, the gel may be viewed through the use of a photo-documentation system.

Silver staining is used when more sensitive method for detection is needed, as classical Coomassie Brilliant Blue staining can usually detect a 50 ng protein band, Silver staining increases the sensitivity typically 10-100 fold more. This is based on the chemistry of photographic development. The proteins are fixed to the gel with a dilute methanol solution, then incubated with an acidic silver nitrate solution. Silver ions are reduced to their metallic form by formaldehyde at alkaline pH. An acidic solution, such as acetic acid stops development. Silver staining was introduced by Kerenyi and Gallyas as a sensitive procedure to detect trace amounts of proteins in gels.

The technique has been extended to the study of other biological macromolecules that have been separated in a variety of supports. Many variables can influence the colour intensity and every protein has its own staining characteristics; clean glassware, pure reagents and water of highest purity are the key points to successful staining. Silver staining was developed in the 14th century for colouring the surface of glass. It has been used extensively for this purpose since the 16th century. The colour produced by the early silver stains ranged between light yellow and an orange-red. Camillo Golgi perfected the silver staining for the study of the nervous system. Golgi's method stains a limited number of cells at random in their entirety.

Autoradiography, also used for protein band detection post gel electrophoresis, uses radioactive isotopes to label proteins, which are then detected by using X-ray film.

Western blotting is a process by which proteins separated in the acrylamide gel are electrophoretically transferred to a stable, manipulable membrane such as a nitrocellulose, nylon, or PVDF membrane. It is then possible to apply immunochemical techniques to visualise the transferred proteins, as well as accurately identify relative increases or decreases of the protein of interest.

### 6.2.3 Capillary Electrophoresis (CE)

Capillary electrophoresis (CE) is a family of electrokinetic separation methods performed in submillimeter diameter capillaries and in micro- and nanofluidic channels. Very often, CE refers to capillary zone electrophoresis (CZE), but other electrophoretic techniques including capillary gel electrophoresis (CGE), capillary isoelectric focusing (CIEF), capillary isotachophoresis and micellar electrokinetic chromatography (MEKC) belong also to this class of methods. In CE methods, analytes migrate through electrolyte solutions under the influence of an electric field. Analytes can be separated according to ionic mobility and/or partitioning into an alternate phase via non-covalent interactions. Additionally, analytes may be concentrated or "focused" by means of gradients in conductivity and pH.

#### Instrumentation

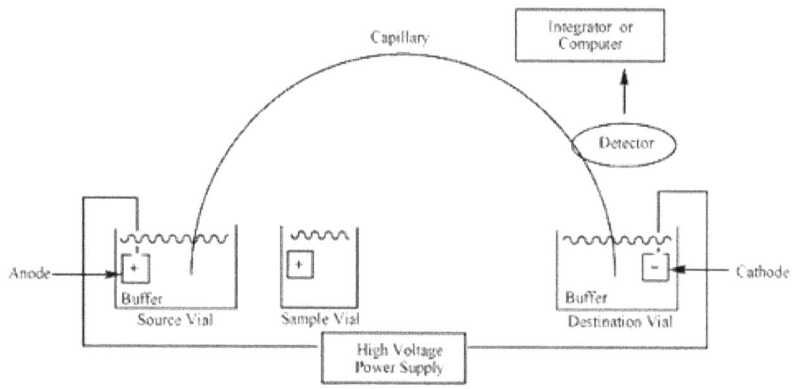


**Fig 1: Diagram of Capillary Electrophoresis System**

The instrumentation needed to perform capillary electrophoresis is relatively simple. The system's main components are a sample vial, source and destination vials, a capillary, electrodes, a high voltage power supply, a detector, and a data output and handling device. The source vial, destination vial and capillary are filled with an electrolyte such as an aqueous buffer solution. To introduce the sample, the capillary inlet is placed into a vial containing the sample. Sample is introduced into the capillary via capillary action, pressure, siphoning, or electrokinetically, and the capillary is then returned to the source vial. The migration of the analytes is initiated by an electric field that is applied between the source and destination vials and is supplied to the electrodes by the high-voltage power supply. In the most common mode of CE, all ions, positive or negative, are pulled through the capillary in the same direction by electroosmotic flow. The analytes separate as they migrate due to their electrophoretic mobility, and are detected near the outlet end of the capillary. The output of the detector is sent to a data output and handling device such as an integrator or computer.

The data is then displayed as an electropherogram, which reports detector response as a function of time. Separated chemical compounds appear as peaks with different migration times in an electropherogram. The technique is often attributed to James W. Jorgensen and Krynn DeArman Lukacs, who first demonstrated the capabilities of this technique. Capillary electrophoresis was first combined with mass spectrometry by Richard D. Smith and coworkers, and provides extremely high sensitivity for the analysis of very small sample sizes. Despite the very small sample sizes (typically only a few nanoliters of liquid are introduced into the capillary), high sensitivity and sharp peaks are achieved in part due to injection strategies that result in a concentration of analytes into a narrow zone near the inlet of the capillary.

This is achieved in either pressure or electrokinetic injections simply by suspending the sample in a buffer of lower conductivity (*e.g.* lower salt concentration) than the running buffer. A process called field-amplified sample stacking (a form of isotachophoresis) results in concentration of analyte in a narrow zone at the boundary between the low-conductivity sample and the higher-conductivity running buffer.

To achieve greater sample throughput, instruments with arrays of capillaries are used to analyze many samples simultaneously. Such capillary array electrophoresis (CAE) instruments with 16 or 96 capillaries are used for medium- to high-throughput capillary DNA sequencing, and the inlet ends of the capillaries are arrayed spatially to accept samples directly from SBS-standard footprint 96-well plates. Certain aspects of the instrumentation (such as detection) are necessarily more complex than for a single-capillary system, but the fundamental principles of design and operation are similar.

## Detection

Separation by capillary electrophoresis can be detected by several detection devices. The majority of commercial systems use UV or UV-Vis absorbance as their primary mode of detection. In these systems, a section of the capillary itself is used as the detection cell. The use of on-tube detection enables detection of separated analytes with no loss of resolution. In general, capillaries used in capillary electrophoresis are coated with a polymer (frequently polyimide or Teflon) for increased flexibility. The portion of the capillary used for UV detection, however, must be optically transparent. For polyimide-coated capillaries, a segment of the coating is typically burned or scraped off to provide a bare window several millimeters long. This bare section of capillary can break easily, and capillaries with transparent coatings are available to increase the stability of the cell window. The path length of the detection cell in capillary electrophoresis (~ 50 micrometers) is far less than that of a traditional UV cell (~ 1 cm). According to the Beer-Lambert law, the sensitivity of the detector is proportional to the path length of the cell. To improve the sensitivity, the path length can be increased, though this results in a loss of resolution.

The capillary tube itself can be expanded at the detection point, creating a "bubble cell" with a longer path length or additional tubing can be added at the detection point as shown in *figure.* Both of these methods, however, will decrease the resolution of the separation. This decrease is almost unnoticeable if a smooth aneurysm is produced in the wall of a capillary by heating and pressurization, as plug flow can be preserved. This invention by Gary Gordon, US Patent 5061361, typically triples the absorbance path length. When used with a UV absorbance detector, the wider cross-section of the analyte in the cell allows for an illuminating beam twice as large, which reduces shot noise by a factor of two. Together these two factors increase the sensitivity of Agilent Technologies's Bubble Cell CE Detector six times over that of one using a straight capillary. This cell and its manufacture are described on page 62 of the June 1995 issue of the *Hewlett-Packard Journal.*

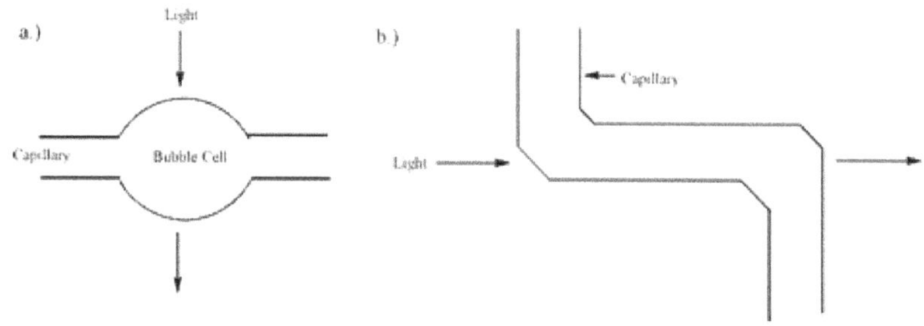


**Fig 2: Techniques for Increasing the Path-Length of the Capillary: a) A Bubble Cell and b) A Z-Cell (Additional Tubing)**

Fluorescence detection can also be used in capillary electrophoresis for samples that naturally fluoresce or are chemically modified to contain fluorescent tags. This mode of detection offers high sensitivity and improved selectivity for these samples, but cannot be utilized for samples that do not fluoresce. Numerous labeling strategies are used to create fluorescent derivatives or conjugates of non-fluorescent molecules, including proteins and DNA.

The set-up for fluorescence detection in a capillary electrophoresis system can be complicated. The method requires that the light beam be focused on the capillary, which can be difficult for many light sources. Laser-induced fluorescence has been used in CE systems with detection limits as low as $10^{-18}$ to $10^{-21}$ mol. The sensitivity of the technique is attributed to the high intensity of the incident light and the ability to accurately focus the light on the capillary. Multi-color fluorescence detection can be achieved by including multiple dichroic mirrors and bandpass filters to separate the fluorescence emission amongst multiple detectors (*e.g.*, photomultiplier tubes), or by using a prism or grating to project spectrally resolved fluorescence emission onto a position-sensitive detector such as a CCD array. CE systems with 4- and 5-color LIF detection systems are used routinely for capillary DNA sequencing and genotyping ("DNA fingerprinting") applications.

In order to obtain the identity of sample components, capillary electrophoresis can be directly coupled with mass spectrometers or surface-enhanced Raman spectroscopy (SERS). In most systems, the capillary outlet is introduced into an ion source that utilizes electrospray ionization (ESI). The resulting ions are then analyzed by the mass spectrometer. This setup requires volatile buffer solutions, which will affect the range of separation modes that can be employed and the degree of resolution that can be achieved. The measurement and analysis are mostly done with a specialized.

For CE-SERS, capillary electrophoresis eluants can be deposited onto a SERS-active substrate. Analyte retention times can be translated into spatial distance by moving the SERS-active substrate at a constant rate during capillary electrophoresis. This allows the subsequent spectroscopic technique to be applied to specific eluants for identification with high sensitivity. SERS-active substrates can be chosen that do not interfere with the spectrum of the analytes.

## Efficiency and Resolution

The number of theoretical plates, or separation efficiency, in capillary electrophoresis is given by:

$$N = \frac{\mu V}{2D_m}$$

According to this equation, the efficiency of separation is only limited by diffusion and is proportional to the strength of the electric field, although practical considerations limit the strength of the electric field to several hundred volts per centimeter. Application of very high potentials (>20-30 kV) may lead to arcing or breakdown of the capillary.

Further, application of strong electric fields leads to resistive heating (Joule heating) of the buffer in the capillary. At sufficiently high field strengths, this heating is strong enough that radial temperature gradients can develop within the capillary. Since electrophoretic mobility of ions is generally temperature-dependent (due to both temperature-dependent ionization and solvent viscosity effects), a non-uniform temperature profile results in variation of electrophoretic mobility across the capillary, and a loss of resolution. The onset of significant Joule heating can be determined by constructing an "Ohm's Law plot", wherein the current through the capillary is measured as a function of applied potential. At low fields, the current is proportional to the applied potential (Ohm's Law), whereas at higher fields the current deviates from the straight line as heating results in decreased resistance of the buffer.

The best resolution is typically obtained at the maximum field strength for which Joule heating is insignificant (*i.e.* near the boundary between the linear and nonlinear regimes of the Ohm's Law plot). Generally capillaries of smaller inner diameter support use of higher field strengths, due to improved heat dissipation and smaller thermal gradients relative to larger capillaries, but with the drawbacks of lower sensitivity in absorbance detection due to shorter path length, and greater difficulty in introducing buffer and sample into the capillary (small capillaries require greater pressure and/or longer times to force fluids through the capillary).

The efficiency of capillary electrophoresis separations is typically much higher than the efficiency of other separation techniques like HPLC. Unlike HPLC, in capillary electrophoresis there is no mass transfer between phases. In addition, the flow profile in EOF-driven systems is flat, rather than the rounded laminar flow profile characteristic of the pressure-driven flow in chromatography columns as shown in *figure*. As a result, EOF does not significantly contribute to band broadening as in pressure-driven chromatography. Capillary electrophoresis separations can have several hundred thousand theoretical plates.

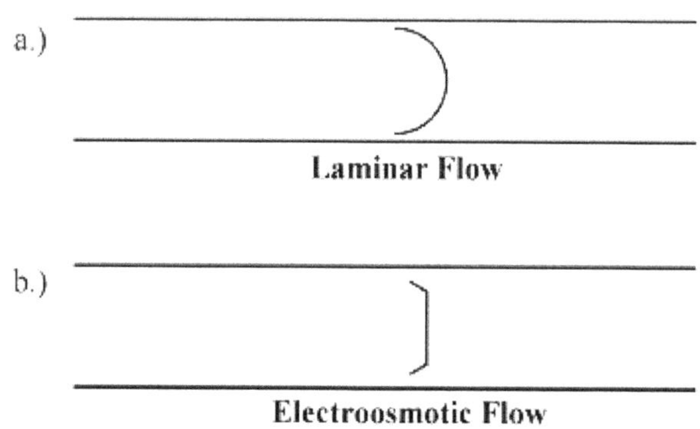


**Fig 3: Flow Profiles of Laminar and Electroosmotic Flow**

The resolution ($R_s$) of capillary electrophoresis separations can be written as:

$$R_s = \frac{1}{4}\left(\frac{\Delta\mu_p\sqrt{N}}{\mu_p + \mu_o}\right)$$

According to this equation, maximum resolution is reached when the electrophoretic and electroosmotic mobilities are similar in magnitude and opposite in sign. In addition, it can be seen that high resolution requires lower velocity and, correspondingly, increased analysis time. Besides diffusion and Joule heating factors that may decrease the resolution in capillary electrophoresis from the theoretical limits in the equation include, but are not limited to, the finite widths of the injection plug and detection window; interactions between the analyte and the capillary wall; instrumental non-idealities such as a slight difference in height of the fluid reservoirs leading to siphoning; irregularities in the electric field due to, *e.g.*, imperfectly cut capillary ends; depletion of buffering capacity in the reservoirs; and electrodispersion (when an analyte has higher conductivity than the background electrolyte). Identifying and minimizing the numerous sources of band broadening is key to successful method development in capillary electrophoresis, with the objective of approaching as close as possible to the ideal of diffusion-limited resolution.

## Applications

Capillary electrophoresis may be used for the simultaneous determination of the ions $NH_4^+$, $Na^+$, $K^+$, $Mg^{2+}$ and $Ca^{2+}$ in saliva. One of the main applications of CE in forensic science is the development of methods for amplification and detection of DNA fragments using polymerase chain reaction (PCR), which has led to rapid and dramatic advances in forensic DNA analysis. DNA separations are carried out using thin CE 50-mm fused silica capillaries filled with a sieving buffer. These capillaries have excellent capabilities to dissipate heat, permitting much higher electric field strengths to be used than slab gel electrophoresis. Therefore separations in capillaries are rapid and efficient. Additionally, the capillaries can be easily refilled and changed for efficient and automated injections. Detection occurs via fluorescence through a window etched in the capillary. Both single-capillary and capillary-array instruments are available with array systems capable of running 16 or more samples simultaneously for increased throughput.

A major use of CE by forensic biologists is typing of STR from biological samples to generate a profile from highly polymorphic genetic markers which differ between individuals. Other emerging uses for CE include the detection of specific mRNA fragments to help identify the biological fluid or tissue origin of a forensic sample.

Another application of CE in forensics is ink analysis, where the analysis of inkjet printing inks is becoming more necessary due to increasingly frequent counterfeiting of documents printed by inkjet printers. The chemical composition of inks provides very important information in cases of fraudulent documents and counterfeit banknotes. Micellar electrophoretic capillary chromatography (MECC) has been developed and applied to the analysis of inks extracted from paper. Due to its high resolving power relative to inks containing several chemically similar substances, differences between inks from the same manufacturer can also be distinguished. This makes it suitable for evaluating the origin of documents based on the chemical composition of inks. It is worth noting that because of the possible compatibility of the same cartridge with different printer models, the differentiation of inks on the basis of their MECC electrophoretic profiles is a more reliable method for the determination of the ink cartridge of origin (its producer and cartridge number) rather than the printer model of origin.

A specialized type of CE, affinity capillary electrophoresis (ACE), utilizes intermolecular binding interactions to understand protein-ligand interactions. Pharmaceutical companies use ACE for a multitude of reasons, with one of the main ones being the association/binding constants for drugs and ligands or drugs and certain vehicle systems like micelles. It is a widely used technique because of its simplicity, rapid results, and low analyte usage. The use of ACE can provide specific details in binding, separation, and detection of analytes and is proven to be highly practical for studies in life sciences. Aptamer-based affinity capillary electrophoresis is utilized for the analysis and modifications of specific affinity reagents. Modified aptamers ideally exhibit and high binding affinity, specificity, and nuclease resistance. Ren et al. incorporated modified nucleotides in aptamers to introduce new confrontational features and high affinity interactions from the hydrophobic and polar interactions between IL-1$\alpha$ and the aptamer. Huang et al. uses ACE to investigate protein-protein interactions using aptamers. A $\alpha$-thrombin binding aptamer was labeled with 6-carboxyfluorescein for use as a selective fluorescent probe and was studied to elucidate information on binding sites for protein-protein and protein-DNA interactions.

Capillary electrophoresis (CE) has become an important, cost-effective approach to do DNA sequencing that provides high throughput and high accuracy sequencing information. Woolley and Mathies used a CE chip to sequence DNA fragments with 97% accuracy and a speed of 150 bases in 540 seconds. They used a 4-color labeling and detection format to collect fluorescent data. Fluorescence is used to view the concentrations of each part of the nucleic acid sequence, A, T, C and G, and these concentration peaks that are graphed from the detection are used to determine the sequence of the DNA.

### 6.2.4 Isoelectric Focusing (IEF)

Isoelectric focusing (IEF), also known as electrofocusing, is a technique for separating different molecules by differences in their isoelectric point (pI). It is a type of zone electrophoresis usually performed on proteins in a gel that takes advantage of the fact that overall charge on the molecule of interest is a function of the pH of its surroundings.

#### Procedure

IEF involves adding an ampholyte solution into immobilized pH gradient (IPG) gels.

IPGs are the acrylamide gel matrix co-polymerized with the pH gradient, which result in completely stable gradients except the most alkaline (>12) pH values. The immobilized pH gradient is obtained by the continuous change in the ratio of *immobilines*. An immobiline is a weak acid or base defined by its pK value.

A protein that is in a pH region below its isoelectric point (pI) will be positively charged and so will migrate toward the cathode (negatively charged electrode). As it migrates through a gradient of increasing pH, however, the protein's overall charge will decrease until the protein reaches the pH region that corresponds to its pI. At this point it has no net charge and so migration ceases (as there is no electrical attraction toward either electrode). As a result, the proteins become focused into sharp stationary bands with each protein positioned at a point in the pH gradient corresponding to its pI. The technique is capable of extremely high resolution with proteins differing by a single charge being fractionated into separate bands.

Molecules to be focused are distributed over a medium that has a pH gradient (usually created by aliphatic ampholytes). An electric current is passed through the medium, creating a "positive" anode and "negative" cathode end. Negatively charged molecules migrate through the pH gradient in the medium toward the "positive" end while positively charged molecules move toward the "negative" end. As a particle moves toward the pole opposite of its charge it moves through the changing pH gradient until it reaches a point in which the pH of that molecule's isoelectric point is reached. At this point the molecule no longer has a net electric charge (due to the protonation or deprotonation of the associated functional groups) and as such will not proceed any further within the gel. The gradient is established before adding the particles of interest by first subjecting a solution of small molecules such as polyampholytes with varying pI values to electrophoresis.

The method is applied particularly often in the study of proteins, which separate based on their relative content of acidic and basic residues, whose value is represented by the pI. Proteins are introduced into an immobilized pH gradient gel composed of polyacrylamide, starch, or agarose where a pH gradient has been established. Gels with large pores are usually used in this process to eliminate any "sieving" effects, or artifacts in the pI caused by differing migration rates for proteins of differing sizes. Isoelectric focusing can resolve proteins that differ in pI value by as little as 0.01. Isoelectric focusing is the first step in two-dimensional gel electrophoresis, in which proteins are first separated by their pI value and then further separated by molecular weight through SDS-PAGE. Isoelectric focusing, on the other hand, is the only step in preparative native PAGE at constant pH.

## Living Cells

According to some opinions, living eukaryotic cells perform isoelectric focusing of proteins in their interior to overcome a limitation of the rate of metabolic reaction by diffusion of enzymes and their reactants, and to regulate the rate of particular biochemical processes. By concentrating the enzymes of particular metabolic pathways into distinct and small regions of its interior, the cell can increase the rate of particular biochemical pathways by several orders of magnitude. By modification of the isoelectric point (pI) of molecules of an enzyme by, e.g., phosphorylation or dephosphorylation, the cell can transfer molecules of the enzyme between different parts of its interior, to switch on or switch off particular biochemical processes.

### Microfluidic Chip Based

Microchip based electrophoresis is a promising alternative to capillary electrophoresis since it has the potential to provide rapid protein analysis, straightforward integration with other microfluidic unit operations, whole channel detection, nitrocellulose films, smaller sample sizes and lower fabrication costs.

### Multi-Junction

The increased demand for faster and easy-to-use protein separation tools has accelerated the evolution of IEF towards in-solution separations. In this context, a multi-junction IEF system was developed to perform fast and gel-free IEF separations. The multi-junction IEF system utilizes a series of vessels with a capillary passing through each vessel. Part of the capillary in each vessel is replaced by a semipermeable membrane. The vessels contain buffer solutions with different pH values, so that a pH gradient is effectively established inside the capillary. The buffer solution in each vessel has an electrical contact with a voltage divider connected to a high-voltage power supply, which establishes an electrical field along the capillary. When a sample (a mixture of peptides or proteins) is injected in the capillary, the presence of the electrical field and the pH gradient separates these molecules according to their isoelectric points. The multi-junction IEF system has been used to separate tryptic peptide mixtures for two-dimensional proteomics and blood plasma proteins from Alzheimer's disease patients for biomarker discovery.

## 6.3 APPLICATIONS OF ELECTROPHORETIC TECHNIQUES

Electrophoretic techniques have numerous applications across various fields, including molecular biology, diagnostics, and forensic science.

- **Molecular Biology and Genetics:** In molecular biology, electrophoresis is indispensable for DNA and RNA analysis. Agarose gel electrophoresis is routinely used to verify PCR products, confirm restriction enzyme digests, and visualize genetic material. Electrophoresis has also contributed significantly to the Human Genome Project and other genomic research initiatives, enabling precise mapping and sequencing of DNA.

- **Proteomics**: In proteomics, electrophoresis, particularly SDS-PAGE and 2D-PAGE, is crucial for studying protein expression, modifications, and interactions. Electrophoresis allows researchers to separate and identify proteins within complex samples, supporting applications in drug discovery, disease biomarker identification, and understanding cellular processes.
- **Clinical Diagnostics**: Electrophoresis is widely used in clinical laboratories to diagnose diseases and assess patient health. Serum protein electrophoresis, for example, helps identify abnormal protein patterns associated with conditions like multiple myeloma and liver disease. Hemoglobin electrophoresis is another diagnostic application, used to identify different types of hemoglobin variants in blood disorders such as sickle cell anemia and thalassemia.
- **Forensic Science**: In forensic science, electrophoresis is used for DNA fingerprinting, which is crucial for identifying individuals in criminal investigations and establishing paternity. Short tandem repeat (STR) analysis, a common technique in forensic DNA profiling, relies on capillary electrophoresis to separate DNA fragments based on size, enabling precise identification of genetic markers.
- **Pharmaceutical Research**: Electrophoresis plays a significant role in the pharmaceutical industry for analyzing drug purity, stability, and composition. Capillary electrophoresis, with its high sensitivity and reproducibility, is particularly useful in quality control processes and pharmacokinetic studies, ensuring the safety and efficacy of pharmaceutical products.

## 6.4 ADVANTAGES AND LIMITATIONS OF ELECTROPHORETIC TECHNIQUES

Electrophoretic techniques offer several advantages, including versatility, simplicity, and high resolution for molecular separation. However, they also have limitations that must be considered.

### 6.4.1 Advantages

- **Versatile:** Electrophoresis can be applied to a wide range of biomolecules, from nucleic acids to proteins and small metabolites.
- **High Resolution:** The ability to separate molecules based on size, charge, or pI provides detailed information, especially in complex mixtures.

- **Relative Simplicity:** Many electrophoretic methods, such as gel electrophoresis, are straightforward and accessible to most research labs.

### 6.4.2 Limitations

- **Sample Size:** Some electrophoretic methods, such as gel electrophoresis, require a relatively large amount of sample, which can be a limitation for rare or difficult-to-purify biomolecules.
- **Sensitivity to Environmental Conditions:** Changes in pH, temperature, and ionic strength can affect separation quality, particularly in protein analysis.
- **Limitations in Resolution:** While electrophoresis offers high resolution, it may not be suitable for separating molecules that are very similar in size and charge, necessitating complementary techniques.

## 6.5 RECENT ADVANCES IN ELECTROPHORETIC TECHNIQUES

Advancements in electrophoresis have led to the development of new methods and improved equipment, enhancing the technique's speed, resolution, and versatility.

### 6.5.1 Microchip Electrophoresis

Microchip electrophoresis, a miniaturized form of electrophoresis, is being used in labs for rapid and high-throughput analysis of DNA, RNA, and proteins. It requires smaller samples and reagents, making it efficient and cost-effective for routine analysis.

### 6.5.2 2D Gel Electrophoresis

Two-dimensional gel electrophoresis, abbreviated as 2-DE or 2-D electrophoresis, is a form of gel electrophoresis commonly used to analyze proteins. Mixtures of proteins are separated by two properties in two dimensions on 2D gels. 2-DE was first independently introduced by O'Farrell and Klose in 1975.

#### Basis for Separation

2-D electrophoresis begins with electrophoresis in the first dimension and then separates the molecules perpendicularly from the first to create an electropherogram in the second dimension.

In electrophoresis in the first dimension, molecules are separated linearly according to their isoelectric point. In the second dimension, the molecules are then separated at 90 degrees from the first electropherogram according to molecular mass. Since it is unlikely that two molecules will be similar in two distinct properties, molecules are more effectively separated in 2-D electrophoresis than in 1-D electrophoresis.

The two dimensions that proteins are separated into using this technique can be isoelectric point, protein complex mass in the native state, or protein mass.

- The separation by isoelectric point is called isoelectric focusing. Thereby, a pH gradient is applied to a gel and an electric potential is applied across the gel, making one end more positive than the other. At all pH values other than their isoelectric point, proteins will be charged. If they are positively charged, they will be pulled towards the more negative end of the gel and if they are negatively charged they will be pulled to the more positive end of the gel. The proteins applied in the first dimension will move along the gel and will accumulate at their isoelectric point; that is, the point at which the overall charge on the protein is 0 (a neutral charge).
- Separation by protein complex mass is done via native PAGE, in which proteins remain in their native state and are separated in the electric field following their mass and the mass of their complexes respectively. To obtain a separation by size and not by net charge, as in IEF, an additional charge is transferred to the proteins by the use of Coomassie brilliant blue or lithium dodecyl sulfate. Knowledge of protein complex is important for the analysis of the functioning of proteins in a cell, as proteins mostly act together in complexes to be fully functional. The analysis of this sub organelle organisation of the cell requires techniques conserving the native state of the protein complexes.
- Separate just by mass is commonly achieved using SDS-PAGE. SDS denatures the proteins, breaks apart most complexes, and approximately equalizes the mass-to-charge ratios. SDS must be done as the second, perpendicular dimension, as it breaks apart complexes (rendering native PAGE impossible) and equalizes mass-to-charge ratios (rendering IEF impossible).

## Detecting Proteins

The result of this is a gel with proteins spread out on its surface. These proteins can then be detected by a variety of means, but the most commonly used stains are silver and Coomassie brilliant blue staining. In the former case, a silver colloid is applied to the gel. The silver binds to cysteine groups within the protein. The silver is darkened by exposure to ultra-violet light. The amount of silver can be related to the darkness, and therefore the amount of protein at a given location on the gel. This measurement can only give approximate amounts, but is adequate for most purposes. Silver staining is 100x more sensitive than Coomassie brilliant blue with a 40-fold range of linearity.

Molecules other than proteins can be separated by 2D electrophoresis. In supercoiling assays, coiled DNA is separated in the first dimension and denatured by a DNA intercalator (such as ethidium bromide or the less carcinogenic chloroquine) in the second. This is comparable to the combination of native PAGE/SDS-PAGE in protein separation.

## Common Techniques

### IPG-DALT

A common technique is to use an Immobilized pH gradient (IPG) in the first dimension. This technique is referred to as IPG-DALT. The sample is first separated onto IPG gel (which is commercially available) then the gel is cut into slices for each sample which is then equilibrated in SDS-mercaptoethanol and applied to an SDS-PAGE gel for resolution in the second dimension. Typically IPG-DALT is not used for quantification of proteins due to the loss of low molecular weight components during the transfer to the SDS-PAGE gel.

## 2D Gel Analysis Software

In quantitative proteomics, these tools primarily analyze bio-markers by quantifying individual proteins, and showing the separation between one or more protein "spots" on a scanned image of a 2-DE gel. Additionally, these tools match spots between gels of similar samples to show, for example, proteomic differences between early and advanced stages of an illness. Software packages include Delta2D (discontinued), ImageMaster (discontinued), Melanie, PDQuest (discontinued), SameSpots and REDFIN – among others. While this technology is widely utilized, the intelligence has not been perfected.

For example, while PDQuest and SameSpots tend to agree on the quantification and analysis of well-defined well-separated protein spots, they deliver different results and analysis tendencies with less-defined less-separated spots. Comparative studies have previously been published to guide researchers on the "best" software for their analysis.

Although typically used for standard gel electrophoresis, Sciugo can also be used for figure-creation and quantification.

Challenges for automatic software-based analysis include incompletely separated (overlapping) spots (less-defined or separated), weak spots / noise (e.g., "ghost spots"), running differences between gels (e.g., protein migrates to different positions on different gels), unmatched/undetected spots, leading to missing values, mismatched spots, errors in quantification (several distinct spots may be erroneously detected as a single spot by the software and parts of a spot may be excluded from quantification), and differences in software algorithms and therefore analysis tendencies. Generated picking lists can be exported from some software packages and used for the automated in-gel digestion of protein spots, and subsequent identification of the proteins by mass spectrometry. Mass spectrometry analysis can identify precise mass measurements along with the sequencing of peptides that range from 1000–4000 atomic mass units. For an overview of the current approach for software analysis of 2DE gel images.

### 6.5.3 Laser-Induced Fluorescence (LIF)

Laser-induced fluorescence (LIF) or laser-stimulated fluorescence (LSF) is a spectroscopic method in which an atom or molecule is excited to a higher energy level by the absorption of laser light followed by spontaneous emission of light. It was first reported by Zare and coworkers in 1968.

LIF is used for studying structure of molecules, detection of selective species and flow visualization and measurements. The wavelength is often selected to be the one at which the species has its largest cross section. The excited species will after some time, usually in the order of few nanoseconds to microseconds, de-excite and emit light at a wavelength longer than the excitation wavelength. This fluorescent light is typically recorded with a photomultiplier tube (PMT) or filtered photodiodes.

#### Types

Two different kinds of spectra exist, disperse spectra and excitation spectra. The disperse spectra are performed with a fixed lasing wavelength. Excitation scans on the other hand collect fluorescent light at a fixed emission wavelength or range of wavelengths. Instead the lasing wavelength is changed.

An advantage over absorption spectroscopy is that it is possible to get two- and three-dimensional images since fluorescence takes place in all directions (i.e. the fluorescence signal is usually isotropic).

The signal-to-noise ratio of the fluorescence signal is very high, providing a good sensitivity to the process. It is also possible to distinguish between more species, since the lasing wavelength can be tuned to a particular excitation of a given species which is not shared by other species.

LIF is useful in the study of the electronic structure of molecules and their interactions. It has also been successfully applied for quantitative measurement of concentrations in fields like combustion, plasma, spray and flow phenomena (such as molecular tagging velocimetry), in some cases visualizing concentrations down to nanomolar levels. LED-induced fluorescence has been used in situ to delineate aromatic hydrocarbon contamination as a cone penetrometer add on module and also as a percussive capable asset.

### Applications

- Detection of purity.
- Optical tumor diagnosis.
- Imaging of paleontological specimens.
- Detection and quantification of biomolecules and biological processes (e.g. DNA sequencing, trace protein analysis, polymerase chain reaction products, and single-cell analysis).
- In plasma diagnostics, which measures plasma properties such as the ion distribution function and velocity space diffusion and convection in a plasma.

## 6.6 CONCLUSION

Electrophoretic techniques are indispensable tools in life sciences research and diagnostics, providing high-resolution separation and analysis of a variety of biomolecules. With advancements in technology and the development of new methods, electrophoresis continues to be a cornerstone in molecular biology, proteomics, and clinical diagnostics. Its versatility, precision, and ability to handle complex samples make it invaluable for studying molecular properties, interactions, and functions. As research demands grow and technology evolves, electrophoretic techniques will continue to advance, supporting discoveries and innovations across scientific disciplines.

# CHAPTER 7
# IMMUNOCHEMICAL TECHNIQUES

Immunochemical techniques are a group of laboratory methods that utilize the highly specific interaction between antigens and antibodies to detect, quantify, or isolate biomolecules. These methods have become indispensable in fields such as immunology, molecular biology, biochemistry, and clinical diagnostics. They are widely used for detecting diseases, identifying proteins, analyzing gene expression, and understanding cellular pathways. Immunochemical techniques include immunoassays, immunohistochemistry, western blotting, ELISA, immunoprecipitation, and flow cytometry, each offering unique applications in research and medicine.

## 7.1 BASIC PRINCIPLES OF IMMUNOCHEMICAL TECHNIQUES

Immunochemical techniques rely on the principle of antigen-antibody binding. Antibodies are proteins produced by the immune system in response to foreign substances, known as antigens. Each antibody has a unique structure that allows it to recognize and bind specifically to its corresponding antigen with high affinity. This specific interaction underlies immunochemical assays, where antibodies are used to detect or quantify target antigens in a sample.

### 7.1.1 Antigens and Antibodies

#### Antigen

In immunology, an antigen (Ag) is a molecule, moiety, foreign particulate matter, or an allergen, such as pollen, that can bind to a specific antibody or T-cell receptor. The presence of antigens in the body may trigger an immune response.

Antigens can be proteins, peptides (amino acid chains), polysaccharides (chains of simple sugars), lipids, or nucleic acids. Antigens exist on normal cells, cancer cells, parasites, viruses, fungi, and bacteria.

Antigens are recognized by antigen receptors, including antibodies and T-cell receptors. Diverse antigen receptors are made by cells of the immune system so that each cell has a specificity for a single antigen. Upon exposure to an antigen, only the lymphocytes that recognize that antigen are activated and expanded, a process known as clonal selection. In most cases, antibodies are *antigen-specific*, meaning that an antibody can only react to and bind one specific antigen; in some instances, however, antibodies may cross-react to bind more than one antigen. The reaction between an antigen and an antibody is called the antigen-antibody reaction.

Antigen can originate either from within the body ("self-protein" or "self antigens") or from the external environment ("non-self"). The immune system identifies and attacks "non-self" external antigens. Antibodies usually do not react with self-antigens due to negative selection of T cells in the thymus and B cells in the bone marrow. The diseases in which antibodies react with self antigens and damage the body's own cells are called autoimmune diseases.

Vaccines are examples of antigens in an immunogenic form, which are intentionally administered to a recipient to induce the memory function of the adaptive immune system towards antigens of the pathogen invading that recipient. The vaccine for seasonal influenza is a common example.

### Exogenous Antigens

Exogenous antigens are antigens that have entered the body from the outside, for example, by inhalation, ingestion or injection. The immune system's response to exogenous antigens is often subclinical. By endocytosis or phagocytosis, exogenous antigens are taken into the antigen-presenting cells (APCs) and processed into fragments. APCs then present the fragments to T helper cells ($CD4^+$) by the use of class II histocompatibility molecules on their surface. Some T cells are specific for the peptide:MHC complex.

They become activated and start to secrete cytokines, substances that activate cytotoxic T lymphocytes (CTL), antibody-secreting B cells, macrophages and other particles. Some antigens start out as exogenous and later become endogenous (for example, intracellular viruses). Intracellular antigens can be returned to circulation upon the destruction of the infected cell.

### Endogenous Antigens

Endogenous antigens are generated within normal cells as a result of normal cell metabolism, or because of viral or intracellular bacterial infection. The fragments are then presented on the cell surface in the complex with MHC class I molecules. If activated cytotoxic $CD8^+$ T cells recognize them, the T cells secrete various toxins that cause the lysis or apoptosis of the infected cell.

In order to keep the cytotoxic cells from killing cells just for presenting self-proteins, the cytotoxic cells (self-reactive T cells) are deleted as a result of tolerance (negative selection). Endogenous antigens include xenogenic (heterologous), autologous and idiotypic or allogenic (homologous) antigens. Sometimes antigens are part of the host itself in an autoimmune disease.

## Autoantigens

An autoantigen is usually a self-protein or protein complex (and sometimes DNA or RNA) that is recognized by the immune system of patients with a specific autoimmune disease. Under normal conditions, these self-proteins should not be the target of the immune system, but in autoimmune diseases, their associated T cells are not deleted and instead attack.

## Neoantigens

Neoantigens are those that are entirely absent from the normal human genome. As compared with nonmutated self-proteins, neoantigens are of relevance to tumor control, as the quality of the T cell pool that is available for these antigens is not affected by central T cell tolerance. Technology to systematically analyze T cell reactivity against neoantigens became available only recently. Neoantigens can be directly detected and quantified.

- **Viral Antigens:** For virus-associated tumors, such as cervical cancer and a subset of head and neck cancers, epitopes derived from viral open reading frames contribute to the pool of neoantigens.
- **Tumor Antigens:** *Tumor antigens* are those antigens that are presented by MHC class I or MHC class II molecules on the surface of tumor cells. Antigens found only on such cells are called tumor-specific antigens (TSAs) and generally result from a tumor-specific mutation. More common are antigens that are presented by tumor cells and normal cells, called tumor-associated antigens (TAAs). Cytotoxic T lymphocytes that recognize these antigens may be able to destroy tumor cells. Tumor antigens can appear on the surface of the tumor in the form of, for example, a mutated receptor, in which case they are recognized by B cells. For human tumors without a viral etiology, novel peptides (neo-epitopes) are created by tumor-specific DNA alterations.
- **Process:** A large fraction of human tumor mutations are effectively patient-specific. Therefore, neoantigens may also be based on individual tumor genomes. Deep-sequencing technologies can identify mutations within the protein-coding part of the genome (the exome) and predict potential neoantigens. In mice models, for all novel protein sequences, potential MHC-binding peptides were predicted. The resulting set of potential neoantigens was used to assess T cell reactivity. Exome-based analyses were exploited in a clinical setting, to assess reactivity in patients treated by either tumor-infiltrating lymphocyte (TIL) cell therapy or checkpoint blockade. Neoantigen identification was successful for multiple experimental model systems and human

malignancies. The false-negative rate of cancer exome sequencing is low—i.e.: the majority of neoantigens occur within exonic sequence with sufficient coverage. However, the vast majority of mutations within expressed genes do not produce neoantigens that are recognized by autologous T cells. As of 2015 mass spectrometry resolution is insufficient to exclude many false positives from the pool of peptides that may be presented by MHC molecules. Instead, algorithms are used to identify the most likely candidates. These algorithms consider factors such as the likelihood of proteasomal processing, transport into the endoplasmic reticulum, affinity for the relevant MHC class I alleles and gene expression or protein translation levels. The majority of human neoantigens identified in unbiased screens display a high predicted MHC binding affinity. Minor histocompatibility antigens, a conceptually similar antigen class are also correctly identified by MHC binding algorithms. Another potential filter examines whether the mutation is expected to improve MHC binding. The nature of the central TCR-exposed residues of MHC-bound peptides is associated with peptide immunogenicity.

### Nativity

A native antigen is an antigen that is not yet processed by an APC to smaller parts. T cells cannot bind native antigens, but require that they be processed by APCs, whereas B cells can be activated by native ones.

### Antigenic Specificity

Antigenic specificity is the ability of the host cells to recognize an antigen specifically as a unique molecular entity and distinguish it from another with exquisite precision. Antigen specificity is due primarily to the side-chain conformations of the antigen. It is measurable and need not be linear or of a rate-limited step or equation. Both T cells and B cells are cellular components of adaptive immunity.

## Antibody

An antibody (Ab) or immunoglobulin (Ig) is a large, Y-shaped protein belonging to the immunoglobulin superfamily which is used by the immune system to identify and neutralize antigens such as bacteria and viruses, including those that cause disease. Antibodies can recognize virtually any size antigen with diverse chemical compositions from molecules. Each antibody recognizes one or more specific antigens. Antigen literally means "antibody generator", as it is the presence of an antigen that drives the formation of an antigen-specific antibody. Each tip of the "Y" of an antibody contains a paratope that specifically binds to one particular epitope on an antigen, allowing the two molecules to bind together with precision.

Using this mechanism, antibodies can effectively "tag" a microbe or an infected cell for attack by other parts of the immune system, or can neutralize it directly (for example, by blocking a part of a virus that is essential for its invasion).

More narrowly, an *antibody* (*Ab*) can refer to the free (secreted) form of these proteins, as opposed to the membrane-bound form found in a B cell receptor. The term *immunoglobulin* can then refer to both forms. Since they are, broadly speaking, the same protein, the terms are often treated as synonymous.

To allow the immune system to recognize millions of different antigens, the antigen-binding sites at both tips of the antibody come in an equally wide variety. The rest of the antibody structure is much less variable; in humans, antibodies occur in five *classes*, sometimes called *isotypes*: IgA, IgD, IgE, IgG, and IgM. Human IgG and IgA antibodies are also divided into discrete subclasses (IgG1, IgG2, IgG3, IgG4; IgA1 and IgA2). The class refers to the functions triggered by the antibody (also known as effector functions), in addition to some other structural features. Antibodies from different classes also differ in where they are released in the body and at what stage of an immune response. Between species, while classes and subclasses of antibodies may be shared (at least in name), their functions and distribution throughout the body may be different. For example, mouse IgG1 is closer to human IgG2 than human IgG1 in terms of its function.

The term humoral immunity is often treated as synonymous with the antibody response, describing the function of the immune system that exists in the body's humors (fluids) in the form of soluble proteins, as distinct from cell-mediated immunity, which generally describes the responses of T cells (especially cytotoxic T cells). In general, antibodies are considered part of the adaptive immune system, though this classification can become complicated. For example, natural IgM, which are made by B-1 lineage cells that have properties more similar to innate immune cells than adaptive, refers to IgM antibodies made independently of an immune response that demonstrate polyreactivity- they recognize multiple distinct (unrelated) antigens. These can work with the complement system in the earliest phases of an immune response to help facilitate clearance of the offending antigen and delivery of the resulting immune complexes to the lymph nodes or spleen for initiation of an immune response. Hence in this capacity, the function of antibodies is more akin to that of innate immunity than adaptive. Nonetheless, in general antibodies are regarded as part of the adaptive immune system because they demonstrate exceptional specificity (with some exception), are produced through genetic rearrangements (rather than being encoded directly in germline), and are a manifestation of immunological memory.

In the course of an immune response, B cells can progressively differentiate into antibody-secreting cells or into memory B cells. Antibody-secreting cells comprise plasmablasts and plasma cells, which differ mainly in the degree to which they secrete antibody, their lifespan, metabolic adaptations, and surface markers. Plasmablasts are rapidly proliferating, short-lived cells produced in the early phases of the immune response (classically described as arising extrafollicularly rather than from the germinal center) which have the potential to differentiate further into plasma cells.

The literature is sloppy at times and often describes plasmablasts as just short-lived plasma cells- formally this is incorrect. Plasma cells, in contrast, do not divide (they are terminally differentiated), and rely on survival niches comprising specific cell types and cytokines to persist. Plasma cells will secrete huge quantities of antibody regardless of whether or not their cognate antigen is present, ensuring that antibody levels to the antigen in question do not fall to 0, provided the plasma cell stays alive.

The rate of antibody secretion, however, can be regulated, for example, by the presence of adjuvant molecules that stimulate the immune response such as TLR ligands. Long-lived plasma cells can live for potentially the entire lifetime of the organism. Classically, the survival niches that house long-lived plasma cells reside in the bone marrow, though it cannot be assumed that any given plasma cell in the bone marrow will be long-lived.

However, other work indicates that survival niches can readily be established within the mucosal tissues- though the classes of antibodies involved show a different hierarchy from those in the bone marrow. B cells can also differentiate into memory B cells which can persist for decades similarly to long-lived plasma cells. These cells can be rapidly recalled in a secondary immune response, undergoing class switching, affinity maturation, and differentiating into antibody-secreting cells.

Antibodies are central to the immune protection elicited by most vaccines and infections (although other components of the immune system certainly participate and for some diseases are considerably more important than antibodies in generating an immune response, e.g. herpes zoster). Durable protection from infections caused by a given microbe – that is, the ability of the microbe to enter the body and begin to replicate (not necessarily to cause disease) – depends on sustained production of large quantities of antibodies, meaning that effective vaccines ideally elicit persistent high levels of antibody, which relies on long-lived plasma cells.

At the same time, many microbes of medical importance have the ability to mutate to escape antibodies elicited by prior infections, and long-lived plasma cells cannot undergo affinity maturation or class switching. This is compensated for through memory B cells: novel variants of a microbe that still retain structural features of previously encountered antigens can elicit memory B cell responses that adapt to those changes. It has been suggested that long-lived plasma cells secrete B cell receptors with higher affinity than those on the surfaces of memory B cells, but findings are not entirely consistent on this point.

## Structure

Antibodies are heavy (~150 kDa) proteins of about 10 nm in size, arranged in three globular regions that roughly form a Y shape.

In humans and most other mammals, an antibody unit consists of four polypeptide chains; two identical *heavy chains* and two identical *light chains* connected by disulfide bonds. Each chain is a series of domains: somewhat similar sequences of about 110 amino acids each. These domains are usually represented in simplified schematics as rectangles. Light chains consist of one variable domain $V_L$ and one constant domain $C_L$, while heavy chains contain one variable domain $V_H$ and three to four constant domains $C_H1$, $C_H2$,...

Structurally an antibody is also partitioned into two antigen-binding fragments (Fab), containing one $V_L$, $V_H$, $C_L$, and $C_H1$ domain each, as well as the crystallisable fragment (Fc), forming the trunk of the Y shape. In between them is a hinge region of the heavy chains, whose flexibility allows antibodies to bind to pairs of epitopes at various distances, to form complexes (dimers, trimers, etc.), and to bind effector molecules more easily.

In an electrophoresis test of blood proteins, antibodies mostly migrate to the last, gamma globulin fraction. Conversely, most gamma-globulins are antibodies, which is why the two terms were historically used as synonyms, as were the symbols Ig and γ. This variant terminology fell out of use due to the correspondence being inexact and due to confusion with γ (gamma) heavy chains which characterize the IgG class of antibodies.

The variable domains can also be referred to as the $F_V$ region. It is the subregion of Fab that binds to an antigen. More specifically, each variable domain contains three *hypervariable regions* – the amino acids seen there vary the most from antibody to antibody. When the protein folds, these regions give rise to three loops of β-strands, localized near one another on the surface of the antibody. These loops are referred to as the complementarity-determining regions (CDRs), since their shape complements that of an antigen. Three CDRs from each of the heavy and light chains together form an antibody-binding site whose shape can be anything from a pocket to which a smaller antigen binds, to a larger surface, to a protrusion that sticks out into a groove in an antigen. Typically though, only a few residues contribute to most of the binding energy.

The existence of two identical antibody-binding sites allows antibody molecules to bind strongly to multivalent antigen (repeating sites such as polysaccharides in bacterial cell walls, or other sites at some distance apart), as well as to form antibody complexes and larger antigen-antibody complexes.

The structures of CDRs have been clustered and classified by Chothia et al. and more recently by North et al. and Nikoloudis et al. However, describing an antibody's binding site using only one single static structure limits the understanding and characterization of the antibody's function and properties. To improve antibody structure prediction and to take the strongly correlated CDR loop and interface movements into account, antibody paratopes should be described as interconverting states in solution with varying probabilities.

In the framework of the immune network theory, CDRs are also called idiotypes. According to immune network theory, the adaptive immune system is regulated by interactions between idiotypes.

The Fc region (the trunk of the Y shape) is composed of constant domains from the heavy chains. Its role is in modulating immune cell activity: it is where effector molecules bind to, triggering various effects after the antibody Fab region binds to an antigen. Effector cells (such as macrophages or natural killer cells) bind via their Fc receptors (FcR) to the Fc region of an antibody, while the complement system is activated by binding the C1q protein complex. IgG or IgM can bind to C1q, but IgA cannot, therefore IgA does not activate the classical complement pathway.

Another role of the Fc region is to selectively distribute different antibody classes across the body. In particular, the neonatal Fc receptor (FcRn) binds to the Fc region of IgG antibodies to transport it across the placenta, from the mother to the fetus. In addition to this, binding to FcRn endows IgG with an exceptionally long half-life relative to other plasma proteins of 3-4 weeks. IgG3 in most cases (depending on allotype) has mutations at the FcRn binding site which lower affinity for FcRn, which are thought to have evolved to limit the highly inflammatory effects of this subclass.

Antibodies are glycoproteins, that is, they have carbohydrates (glycans) added to conserved amino acid residues. These conserved glycosylation sites occur in the Fc region and influence interactions with effector molecules.

The N-terminus of each chain is situated at the tip. Each immunoglobulin domain has a similar structure, characteristic of all the members of the immunoglobulin superfamily: it is composed of between 7 (for constant domains) and 9 (for variable domains) β-strands, forming two beta sheets in a Greek key motif. The sheets create a "sandwich" shape, the immunoglobulin fold, held together by a disulfide bond.

Secreted antibodies can occur as a single Y-shaped unit, a monomer. However, some antibody classes also form dimers with two Ig units (as with IgA), tetramers with four Ig units (like teleost fish IgM), or pentamers with five Ig units (like shark

IgW or mammalian IgM, which occasionally forms hexamers as well, with six units). IgG can also form hexamers, though no J chain is required. IgA tetramers and pentamers have also been reported.

Antibodies also form complexes by binding to antigen: this is called an antigen-antibody complex or *immune complex*. Small antigens can cross-link two antibodies, also leading to the formation of antibody dimers, trimers, tetramers, etc. Multivalent antigens (e.g., cells with multiple epitopes) can form larger complexes with antibodies. An extreme example is the clumping, or agglutination, of red blood cells with antibodies in blood typing to determine blood groups: the large clumps become insoluble, leading to visually apparent precipitation.

The membrane-bound form of an antibody may be called a *surface immunoglobulin* (sIg) or a *membrane immunoglobulin* (mIg). It is part of the *B cell receptor* (BCR), which allows a B cell to detect when a specific antigen is present in the body and triggers B cell activation. The BCR is composed of surface-bound IgD or IgM antibodies and associated Ig-α and Ig-β heterodimers, which are capable of signal transduction. A typical human B cell will have 50,000 to 100,000 antibodies bound to its surface. Upon antigen binding, they cluster in large patches, which can exceed 1 micrometer in diameter, on lipid rafts that isolate the BCRs from most other cell signalling receptors. These patches may improve the efficiency of the cellular immune response. In humans, the cell surface is bare around the B cell receptors for several hundred nanometers, which further isolates the BCRs from competing influences.

## Classes

Antibodies can come in different varieties known as *isotypes* or *classes*. In humans there are five antibody classes known as IgA, IgD, IgE, IgG, and IgM, which are further subdivided into subclasses such as IgA1, IgA2. The prefix "Ig" stands for *immunoglobulin*, while the suffix denotes the type of heavy chain the antibody contains: the heavy chain types α (alpha), γ (gamma), δ (delta), ε (epsilon), μ (mu) give rise to IgA, IgG, IgD, IgE, IgM, respectively.

The distinctive features of each class are determined by the part of the heavy chain within the hinge and Fc region.

The classes differ in their biological properties, functional locations and ability to deal with different antigens, as depicted in the table. For example, IgE antibodies are responsible for an allergic response consisting of histamine release from mast cells, often a sole contributor to asthma (though other pathways exist as do exist symptoms very similar to yet not technically asthma). The antibody's variable region binds to allergic antigen, for example house dust mite particles, while its Fc region (in the ε heavy chains) binds to Fc receptor ε on a mast cell, triggering its degranulation: the release of molecules stored in its granules.

**Table 1: Antibody Isotypes of Humans**

| Class | Subclasses | Description |
|---|---|---|
| IgA | 2 | Found in mucosal areas, such as the gut, respiratory tract and urogenital tract, and prevents colonization by pathogens. Also found in saliva, tears, and breast milk. Early clinical studies suggest that IgA isotype antibodies have potential as anti-cancer therapeutics, demonstrating the ability to reduce tumor growth. |
| IgD | 1 | Functions mainly as an antigen receptor on B cells that have not been exposed to antigens. It has been shown to activate basophils and mast cells to produce antimicrobial factors. Besides, IgD has also been reported to induce the release of immunoactivity and pro-inflammatory mediators. |
| IgE | 1 | IgE antibodies are the least abundant class of immunoglobulin, they can engage Fc receptors on monocytes and macrophages to activate various effector cell populations.<br>Binds to allergens and triggers histamine release from mast cells and basophils, and is involved in allergy. Humans and other animals evolved IgE to protect against parasitic worms, though in the present, IgE is primarily related to allergies and asthma. Although |
| IgG | 4 | In its four forms, provides the majority of antibody-based immunity against invading pathogens. The only antibody capable of crossing the placenta to give passive immunity to the fetus. IgG is the most commonly used molecular format in current antibody drugs because it neutralizes infectious agents and activates the complement system to engage immune cells. |
| IgM | 1 | Expressed on the surface of B cells (monomer) and in a secreted form (pentamer) with very high avidity. Eliminates pathogens in the early stages of B cell-mediated (humoral) immunity before there is sufficient IgG. IgM also is a pro-inflammatory antibody that serves as the primary defense and effectively stimulates the complement system with specialized immune functions, including higher avidity and steric hindrance, allowing it to neutralize viruses. |

The antibody isotype of a B cell changes during cell development and activation. Immature B cells, which have never been exposed to an antigen, express only the IgM isotype in a cell surface bound form. The B lymphocyte, in this ready-to-respond form, is known as a "naive B lymphocyte." The naive B lymphocyte expresses both surface IgM and IgD.

The co-expression of both of these immunoglobulin isotypes renders the B cell ready to respond to antigen. B cell activation follows engagement of the cell-bound antibody molecule with an antigen, causing the cell to divide and differentiate into an antibody-producing cell called a plasma cell.

In this activated form, the B cell starts to produce antibody in a secreted form rather than a membrane-bound form. Some daughter cells of the activated B cells undergo isotype switching, a mechanism that causes the production of antibodies to change from IgM or IgD to the other antibody isotypes, IgE, IgA, or IgG, that have defined roles in the immune system.

In mammals there are two types of immunoglobulin light chain, which are called lambda (λ) and kappa (κ). However, there is no known functional difference between them, and both can occur with any of the five major types of heavy chains. Each antibody contains two identical light chains: both κ or both λ. Proportions of κ and λ types vary by species and can be used to detect abnormal proliferation of B cell clones. Other types of light chains, such as the iota (ι) chain, are found in other vertebrates like sharks (Chondrichthyes) and bony fishes (Teleostei).

In most placental mammals, the structure of antibodies is generally the same. Jawed fish appear to be the most primitive animals that are able to make antibodies similar to those of mammals, although many features of their adaptive immunity appeared somewhat earlier.

Cartilaginous fish (such as sharks) produce heavy-chain-only antibodies (i.e., lacking light chains) which moreover feature longer chain pentamers (with five constant units per molecule). Camelids (such as camels, llamas, alpacas) are also notable for producing heavy-chain-only antibodies.

### Antibody–Antigen Interactions

The antibody's paratope interacts with the antigen's epitope. An antigen usually contains different epitopes along its surface arranged discontinuously, and dominant epitopes on a given antigen are called determinants. Antibody and antigen interact by spatial complementarity (lock and key). The molecular forces involved in the Fab-epitope interaction are weak and non-specific – for example electrostatic forces, hydrogen bonds, hydrophobic interactions, and van der Waals forces.

This means binding between antibody and antigen is reversible, and the antibody's affinity towards an antigen is relative rather than absolute. Relatively weak binding also means it is possible for an antibody to cross-react with different antigens of different relative affinities.

### Function

The main categories of antibody action include the following:

- **Neutralisation**, in which neutralizing antibodies block parts of the surface of a bacterial cell or virion to render its attack ineffective.
- **Agglutination**, in which antibodies "glue together" foreign cells into clumps that are attractive targets for phagocytosis.
- **Precipitation**, in which antibodies "glue together" serum-soluble

antigens, forcing them to precipitate out of solution in clumps that are attractive targets for phagocytosis.

- **Complement activation (fixation)**, in which antibodies that are latched onto a foreign cell encourage complement to attack it with a membrane attack complex, which leads to the following:
  - Lysis of the foreign cell.
  - Encouragement of inflammation by chemotactically attracting inflammatory cells.

More indirectly, an antibody can signal immune cells to present antibody fragments to T cells, or downregulate other immune cells to avoid autoimmunity.

Activated B cells differentiate into either antibody-producing cells called plasma cells that secrete soluble antibody or memory cells that survive in the body for years afterward in order to allow the immune system to remember an antigen and respond faster upon future exposures.

At the prenatal and neonatal stages of life, the presence of antibodies is provided by passive immunization from the mother. Early endogenous antibody production varies for different kinds of antibodies, and usually appear within the first years of life. Since antibodies exist freely in the bloodstream, they are said to be part of the humoral immune system. Circulating antibodies are produced by clonal B cells that specifically respond to only one antigen (an example is a virus capsid protein fragment). Antibodies contribute to immunity in three ways: They prevent pathogens from entering or damaging cells by binding to them; they stimulate removal of pathogens by macrophages and other cells by coating the pathogen; and they trigger destruction of pathogens by stimulating other immune responses such as the complement pathway. Antibodies will also trigger vasoactive amine degranulation to contribute to immunity against certain types of antigens (helminths, allergens).

Antibodies that bind to surface antigens (for example, on bacteria) will attract the first component of the complement cascade with their Fc region and initiate activation of the "classical" complement system. This results in the killing of bacteria in two ways. First, the binding of the antibody and complement molecules marks the microbe for ingestion by phagocytes in a process called opsonization; these phagocytes are attracted by certain complement molecules generated in the complement cascade. Second, some complement system components form a membrane attack complex to assist antibodies to kill the bacterium directly (bacteriolysis).

To combat pathogens that replicate outside cells, antibodies bind to pathogens to link them together, causing them to agglutinate. Since an antibody has at least two paratopes, it can bind more than one antigen by binding identical epitopes carried on the surfaces of these antigens. By coating the pathogen, antibodies stimulate effector functions against the pathogen in cells that recognize their Fc region.

Those cells that recognize coated pathogens have Fc receptors, which, as the name suggests, interact with the Fc region of IgA, IgG, and IgE antibodies. The engagement of a particular antibody with the Fc receptor on a particular cell triggers an effector function of that cell; phagocytes will phagocytose, mast cells and neutrophils will degranulate, natural killer cells will release cytokines and cytotoxic molecules; that will ultimately result in destruction of the invading microbe.

The activation of natural killer cells by antibodies initiates a cytotoxic mechanism known as antibody-dependent cell-mediated cytotoxicity (ADCC) – this process may explain the efficacy of monoclonal antibodies used in biological therapies against cancer. The Fc receptors are isotype-specific, which gives greater flexibility to the immune system, invoking only the appropriate immune mechanisms for distinct pathogens.

Humans and higher primates also produce "natural antibodies" that are present in serum before viral infection. Natural antibodies have been defined as antibodies that are produced without any previous infection, vaccination, other foreign antigen exposure or passive immunization. These antibodies can activate the classical complement pathway leading to lysis of enveloped virus particles long before the adaptive immune response is activated. Antibodies are produced exclusively by B cells in response to antigens where initially, antibodies are formed as membrane-bound receptors, but upon activation by antigens and helper T cells, B cells differentiate to produce soluble antibodies. Many natural antibodies are directed against the disaccharide galactose $\alpha$(1,3)-galactose ($\alpha$-Gal), which is found as a terminal sugar on glycosylated cell surface proteins, and generated in response to production of this sugar by bacteria contained in the human gut.

These antibodies undergo quality checks in the endoplasmic reticulum (ER), which contains proteins that assist in proper folding and assembly. Rejection of xenotransplantated organs is thought to be, in part, the result of natural antibodies circulating in the serum of the recipient binding to $\alpha$-Gal antigens expressed on the donor tissue.

### Immunoglobulin Diversity

Virtually all microbes can trigger an antibody response. Successful recognition and eradication of many different types of microbes requires diversity among antibodies; their amino acid composition varies allowing them to interact with many different antigens. It has been estimated that humans generate about 10 billion different antibodies, each capable of binding a distinct epitope of an antigen. Although a huge repertoire of different antibodies is generated in a single individual, the number of genes available to make these proteins is limited by the size of the human genome. Several complex genetic mechanisms have evolved that allow vertebrate B cells to generate a diverse pool of antibodies from a relatively small number of antibody genes.

The chromosomal region that encodes an antibody is large and contains several distinct gene loci for each domain of the antibody—the chromosome region containing heavy chain genes (IGH@) is found on chromosome 14, and the loci containing lambda and kappa light chain genes (IGL@ and IGK@) are found on chromosomes 22 and 2 in humans. One of these domains is called the variable domain, which is present in each heavy and light chain of every antibody, but can differ in different antibodies generated from distinct B cells. Differences between the variable domains are located on three loops known as hypervariable regions (HV-1, HV-2 and HV-3) or complementarity-determining regions (CDR1, CDR2 and CDR3). CDRs are supported within the variable domains by conserved framework regions. The heavy chain locus contains about 65 different variable domain genes that all differ in their CDRs. Combining these genes with an array of genes for other domains of the antibody generates a large cavalry of antibodies with a high degree of variability.

Somatic recombination of immunoglobulins, also known as *V(D)J recombination*, involves the generation of a unique immunoglobulin variable region. The variable region of each immunoglobulin heavy or light chain is encoded in several pieces—known as gene segments (subgenes). These segments are called variable (V), diversity (D) and joining (J) segments. V, D and J segments are found in Ig heavy chains, but only V and J segments are found in Ig light chains. Multiple copies of the V, D and J gene segments exist, and are tandemly arranged in the genomes of mammals. In the bone marrow, each developing B cell will assemble an immunoglobulin variable region by randomly selecting and combining one V, one D and one J gene segment (or one V and one J segment in the light chain).

As there are multiple copies of each type of gene segment, and different combinations of gene segments can be used to generate each immunoglobulin variable region, this process generates a huge number of antibodies, each with different paratopes, and thus different antigen specificities. The rearrangement of several subgenes (i.e. V2 family) for lambda light chain immunoglobulin is coupled with the activation of microRNA miR-650, which further influences biology of B-cells.

RAG proteins play an important role with V(D)J recombination in cutting DNA at a particular region. Without the presence of these proteins, V(D)J recombination would not occur. After a B cell produces a functional immunoglobulin gene during V(D)J recombination, it cannot express any other variable region (a process known as allelic exclusion) thus each B cell can produce antibodies containing only one kind of variable chain.

Following activation with antigen, B cells begin to proliferate rapidly. In these rapidly dividing cells, the genes encoding the variable domains of the heavy and light chains undergo a high rate of point mutation, by a process called *somatic hypermutation* (SHM). SHM results in approximately one nucleotide change per variable gene, per cell division. As a consequence, any daughter B cells will acquire slight amino acid differences in the variable domains of their antibody chains.

This serves to increase the diversity of the antibody pool and impacts the antibody's antigen-binding affinity. Some point mutations will result in the production of antibodies that have a weaker interaction (low affinity) with their antigen than the original antibody, and some mutations will generate antibodies with a stronger interaction (high affinity). B cells that express high affinity antibodies on their surface will receive a strong survival signal during interactions with other cells, whereas those with low affinity antibodies will not, and will die by apoptosis. Thus, B cells expressing antibodies with a higher affinity for the antigen will outcompete those with weaker affinities for function and survival allowing the average affinity of antibodies to increase over time. The process of generating antibodies with increased binding affinities is called *affinity maturation*. Affinity maturation occurs in mature B cells after V(D)J recombination, and is dependent on help from helper T cells.

Isotype or class switching is a biological process occurring after activation of the B cell, which allows the cell to produce different classes of antibody (IgA, IgE, or IgG). The different classes of antibody, and thus effector functions, are defined by the constant (C) regions of the immunoglobulin heavy chain. Initially, naive B cells express only cell-surface IgM and IgD with identical antigen binding regions. Each isotype is adapted for a distinct function; therefore, after activation, an antibody with an IgG, IgA, or IgE effector function might be required to effectively eliminate an antigen. Class switching allows different daughter cells from the same activated B cell to produce antibodies of different isotypes. Only the constant region of the antibody heavy chain changes during class switching; the variable regions, and therefore antigen specificity, remain unchanged. Thus the progeny of a single B cell can produce antibodies, all specific for the same antigen, but with the ability to produce the effector function appropriate for each antigenic challenge. Class switching is triggered by cytokines; the isotype generated depends on which cytokines are present in the B cell environment.

Class switching occurs in the heavy chain gene locus by a mechanism called class switch recombination (CSR). This mechanism relies on conserved nucleotide motifs, called *switch (S) regions*, found in DNA upstream of each constant region gene (except in the δ-chain). The DNA strand is broken by the activity of a series of enzymes at two selected S-regions. The variable domain exon is rejoined through a process called non-homologous end joining (NHEJ) to the desired constant region (γ, α or ε). This process results in an immunoglobulin gene that encodes an antibody of a different isotype.

An antibody can be called *monospecific* if it has specificity for a single antigen or epitope, or bispecific if it has affinity for two different antigens or two different epitopes on the same antigen. A group of antibodies can be called *polyvalent* (or *unspecific*) if they have affinity for various antigens or microorganisms. Intravenous immunoglobulin, if not otherwise noted, consists of a variety of different IgG (polyclonal IgG). In contrast, monoclonal antibodies are identical antibodies produced by a single B cell.

Heterodimeric antibodies, which are also asymmetrical antibodies, allow for greater flexibility and new formats for attaching a variety of drugs to the antibody arms. One of the general formats for a heterodimeric antibody is the "knobs-into-holes" format. This format is specific to the heavy chain part of the constant region in antibodies. The "knobs" part is engineered by replacing a small amino acid with a larger one. It fits into the "hole", which is engineered by replacing a large amino acid with a smaller one. What connects the "knobs" to the "holes" are the disulfide bonds between each chain.

The "knobs-into-holes" shape facilitates antibody dependent cell mediated cytotoxicity. Single-chain variable fragments (scFv) are connected to the variable domain of the heavy and light chain via a short linker peptide. The linker is rich in glycine, which gives it more flexibility, and serine/threonine, which gives it specificity. Two different scFv fragments can be connected together, via a hinge region, to the constant domain of the heavy chain or the constant domain of the light chain. This gives the antibody bispecificity, allowing for the binding specificities of two different antigens. The "knobs-into-holes" format enhances heterodimer formation but does not suppress homodimer formation.

To further improve the function of heterodimeric antibodies, many scientists are looking towards artificial constructs. Artificial antibodies are largely diverse protein motifs that use the functional strategy of the antibody molecule, but are not limited by the loop and framework structural constraints of the natural antibody. Being able to control the combinational design of the sequence and three-dimensional space could transcend the natural design and allow for the attachment of different combinations of drugs to the arms.

Heterodimeric antibodies have a greater range in shapes they can take and the drugs that are attached to the arms do not have to be the same on each arm, allowing for different combinations of drugs to be used in cancer treatment. Pharmaceuticals are able to produce highly functional bispecific, and even multispecific, antibodies. The degree to which they can function is impressive given that such a change of shape from the natural form should lead to decreased functionality.

Antibody diversification typically occurs through somatic hypermutation, class switching, and affinity maturation targeting the BCR gene loci, but on occasion more unconventional forms of diversification have been documented. For example, in the case of malaria caused by *Plasmodium falciparum,* some antibodies from those who had been infected demonstrated an insertion from chromosome 19 containing a 98-amino acid stretch from leukocyte-associated immunoglobulin-like receptor 1, LAIR1, in the elbow joint.

This represents a form of interchromosomal transposition. LAIR1 normally binds collagen, but can recognize repetitive interspersed families of polypeptides (RIFIN) family members that are highly expressed on the surface of *P. falciparum*-infected red blood cells. In fact, these antibodies underwent affinity maturation that enhanced affinity for RIFIN but abolished affinity for collagen. These "LAIR1-containing" antibodies have been found in 5-10% of donors from Tanzania and Mali, though not in European donors. European donors did show 100-1000 nucleotide stretches inside the elbow joints as well, however. This particular phenomenon may be specific to malaria, as infection is known to induce genomic instability.

## Medical Applications

- **Disease Diagnosis:**
  - Detection of particular antibodies is a very common form of medical diagnostics, and applications such as serology depend on these methods. For example, in biochemical assays for disease diagnosis, a titer of antibodies directed against Epstein-Barr virus or Lyme disease is estimated from the blood. If those antibodies are not present, either the person is not infected or the infection occurred a *very* long time ago, and the B cells generating these specific antibodies have naturally decayed.
  - In clinical immunology, levels of individual classes of immunoglobulins are measured by nephelometry (or turbidimetry) to characterize the antibody profile of patient. Elevations in different classes of immunoglobulins are sometimes useful in determining the cause of liver damage in patients for whom the diagnosis is unclear. For example, elevated IgA indicates alcoholic cirrhosis, elevated IgM indicates viral hepatitis and primary biliary cirrhosis, while IgG is elevated in viral hepatitis, autoimmune hepatitis and cirrhosis.
  - Autoimmune disorders can often be traced to antibodies that bind the body's own epitopes; many can be detected through blood tests. Antibodies directed against red blood cell surface antigens in immune mediated hemolytic anemia are detected with the Coombs test. The Coombs test is

also used for antibody screening in blood transfusion preparation and also for antibody screening in antenatal women.

- Practically, several immunodiagnostic methods based on detection of complex antigen-antibody are used to diagnose infectious diseases, for example ELISA, immunofluorescence, Western blot, immunodiffusion, immunoelectrophoresis, and magnetic immunoassay. Antibodies raised against human chorionic gonadotropin are used in over the counter pregnancy tests.
- New dioxaborolane chemistry enables radioactive fluoride ($^{18}F$) labeling of antibodies, which allows for positron emission tomography (PET) imaging of cancer.

- **Disease Therapy:**
  - Targeted monoclonal antibody therapy is employed to treat diseases such as rheumatoid arthritis, multiple sclerosis, psoriasis, and many forms of cancer including non-Hodgkin's lymphoma, colorectal cancer, head and neck cancer and breast cancer.
  - Some immune deficiencies, such as X-linked agammaglobulinemia and hypogammaglobulinemia, result in partial or complete lack of antibodies. These diseases are often treated by inducing a short-term form of immunity called passive immunity. Passive immunity is achieved through the transfer of ready-made antibodies in the form of human or animal serum, pooled immunoglobulin or monoclonal antibodies, into the affected individual.
- **Prenatal Therapy:**
  - Rh factor, also known as Rh D antigen, is an antigen found on red blood cells; individuals that are Rh-positive (Rh+) have this antigen on their red blood cells and individuals that are Rh-negative (Rh–) do not. During normal childbirth, delivery trauma or complications during pregnancy, blood from a fetus can enter the mother's system. In the case of an Rh-incompatible mother and child, consequential blood mixing may sensitize an Rh- mother to the Rh antigen on the blood cells of the Rh+ child, putting the remainder of the pregnancy, and any subsequent pregnancies, at risk for hemolytic disease of the newborn.
  - Rho(D) immune globulin antibodies are specific for human RhD antigen. Anti-RhD antibodies are administered as part of a prenatal treatment regimen to prevent sensitization that may occur when a Rh-negative mother has a Rh-positive fetus. Treatment of a mother with Anti-RhD antibodies prior to and immediately after trauma and delivery destroys Rh antigen in the mother's system from the fetus. This occurs before the antigen can stimulate maternal B cells to "remember" Rh antigen by generating memory B cells. Therefore, her humoral immune system will not make anti-Rh antibodies, and will not attack the Rh antigens of the current or subsequent babies. Rho(D) Immune Globulin treatment prevents sensitization that can lead to Rh disease, but does not prevent or treat the underlying disease itself.

## Research Applications

Specific antibodies are produced by injecting an antigen into a mammal, such as a mouse, rat, rabbit, goat, sheep, or horse for large quantities of antibody. Blood isolated from these animals contains *polyclonal antibodies*—multiple antibodies that bind to the same antigen—in the serum, which can now be called antiserum. Antigens are also injected into chickens for generation of polyclonal antibodies in egg yolk.

To obtain antibody that is specific for a single epitope of an antigen, antibody-secreting lymphocytes are isolated from the animal and immortalized by fusing them with a cancer cell line. The fused cells are called hybridomas, and will continually grow and secrete antibody in culture. Single hybridoma cells are isolated by dilution cloning to generate cell clones that all produce the same antibody; these antibodies are called *monoclonal antibodies*. Polyclonal and monoclonal antibodies are often purified using Protein A/G or antigen-affinity chromatography.

In research, purified antibodies are used in many applications. Antibodies for research applications can be found directly from antibody suppliers, or through use of a specialist search engine. Research antibodies are most commonly used to identify and locate intracellular and extracellular proteins.

Antibodies are used in flow cytometry to differentiate cell types by the proteins they express; different types of cells express different combinations of cluster of differentiation molecules on their surface, and produce different intracellular and secretable proteins. They are also used in immunoprecipitation to separate proteins and anything bound to them (co-immunoprecipitation) from other molecules in a cell lysate, in Western blot analyses to identify proteins separated by electrophoresis, and in immunohistochemistry or immunofluorescence to examine protein expression in tissue sections or to locate proteins within cells with the assistance of a microscope. Proteins can also be detected and quantified with antibodies, using ELISA and ELISpot techniques.

Antibodies used in research are some of the most powerful, yet most problematic reagents with a tremendous number of factors that must be controlled in any experiment including cross reactivity, or the antibody recognizing multiple epitopes and affinity, which can vary widely depending on experimental conditions such as pH, solvent, state of tissue etc. Multiple attempts have been made to improve both the way that researchers validate antibodies and ways in which they report on antibodies. Researchers using antibodies in their work need to record them correctly in order to allow their research to be reproducible (and therefore tested, and qualified by other researchers). Less than half of research antibodies referenced in academic papers can be easily identified. Papers published in F1000 in 2014 and 2015 provide researchers with a guide for reporting research antibody use. The

RRID paper, is co-published in 4 journals that implemented the RRIDs Standard for research resource citation, which draws data from the antibodyregistry.org as the source of antibody identifiers.

Antibody regions can be used to further biomedical research by acting as a guide for drugs to reach their target. Several application involve using bacterial plasmids to tag plasmids with the Fc region of the antibody such as pFUSE-Fc plasmid.

## Regulations

There are several ways to obtain antibodies, including in vivo techniques like animal immunization and various in vitro approaches, such as the phage display method. Traditionally, most antibodies are produced by hybridoma cell lines through immortalization of antibody-producing cells by chemically induced fusion with myeloma cells. In some cases, additional fusions with other lines have created "triomas" and "quadromas". The manufacturing process should be appropriately described and validated. Validation studies should at least include:

- The demonstration that the process is able to produce in good quality (the process should be validated).
- The efficiency of the antibody purification (all impurities and virus must be eliminated).
- The characterization of purified antibody (physicochemical characterization, immunological properties, biological activities, contaminants,...).
- Determination of the virus clearance studies.
- **Before Clinical Trials:**
  - *Product Safety Testing:* Sterility (bacteria and fungi), in vitro and in vivo testing for adventitious viruses, murine retrovirus testing..., product safety data needed before the initiation of feasibility trials in serious or immediately life-threatening conditions, it serves to evaluate dangerous potential of the product.
  - *Feasibility Testing:* These are pilot studies whose objectives include, among others, early characterization of safety and initial proof of concept in a small specific patient population (in vitro or in vivo testing).
- **Preclinical Studies:**
  - Testing cross-reactivity of antibody: to highlight unwanted interactions (toxicity) of antibodies with previously characterized tissues. This study can be performed in vitro (reactivity of the antibody or immunoconjugate should be determined with a quick-frozen adult tissues) or in vivo (with appropriates animal models).
  - Preclinical pharmacology and toxicity testing: preclinical safety testing of

antibody is designed to identify possible toxicity in humans, to estimate the likelihood and severity of potential adverse events in humans, and to identify a safe starting dose and dose escalation, when possible.

- Animal toxicity studies: Acute toxicity testing, repeat-dose toxicity testing, long-term toxicity testing
- Pharmacokinetics and pharmacodynamics testing: Use for determinate clinical dosages, antibody activities, evaluation of the potential clinical effects

## Structure Prediction and Computational Antibody Design

The importance of antibodies in health care and the biotechnology industry demands knowledge of their structures at high resolution. This information is used for protein engineering, modifying the antigen binding affinity, and identifying an epitope, of a given antibody. X-ray crystallography is one commonly used method for determining antibody structures. However, crystallizing an antibody is often laborious and time-consuming.

Computational approaches provide a cheaper and faster alternative to crystallography, but their results are more equivocal, since they do not produce empirical structures. Online web servers such as *Web Antibody Modeling* (WAM) and *Prediction of Immunoglobulin Structure* (PIGS) enable computational modeling of antibody variable regions. Rosetta Antibody is a novel antibody $F_V$ region structure prediction server, which incorporates sophisticated techniques to minimize CDR loops and optimize the relative orientation of the light and heavy chains, as well as homology models that predict successful docking of antibodies with their unique antigen.

However, describing an antibody's binding site using only one single static structure limits the understanding and characterization of the antibody's function and properties. To improve antibody structure prediction and to take the strongly correlated CDR loop and interface movements into account, antibody paratopes should be described as interconverting states in solution with varying probabilities.

The ability to describe the antibody through binding affinity to the antigen is supplemented by information on antibody structure and amino acid sequences for the purpose of patent claims. Several methods have been presented for computational design of antibodies based on the structural bioinformatics studies of antibody CDRs.

There are a variety of methods used to sequence an antibody including Edman degradation, cDNA, etc.; albeit one of the most common modern uses for peptide/protein identification is liquid chromatography coupled with tandem mass spectrometry (LC-MS/MS). High volume antibody sequencing methods require computational approaches for the data analysis, including de novo sequencing directly from tandem mass spectra and database search methods that use existing protein sequence databases. Many versions of shotgun protein sequencing are able to increase the coverage by utilizing CID/HCD/ETD fragmentation methods and other techniques, and they have achieved substantial progress in attempt to fully sequence proteins, especially antibodies. Other methods have assumed the existence of similar proteins, a known genome sequence, or combined top-down and bottom up approaches.

Current technologies have the ability to assemble protein sequences with high accuracy by integrating de novo sequencing peptides, intensity, and positional confidence scores from database and homology searches.

#### Antibody Mimetic

Antibody mimetics are organic compounds, like antibodies, that can specifically bind antigens. They consist of artificial peptides or proteins, or aptamer-based nucleic acid molecules with a molar mass of about 3 to 20 kDa. Antibody fragments, such as Fab and nanobodies are not considered as antibody mimetics. Common advantages over antibodies are better solubility, tissue penetration, stability towards heat and enzymes, and comparatively low production costs. Antibody mimetics have being developed and commercialized as research, diagnostic and therapeutic agents.

#### Binding Antibody Unit

BAU (binding antibody unit, often as BAU/mL) is a measurement unit defined by the WHO for the comparison of assays detecting the same class of immunoglobulins with the same specificity.

### 7.1.2 Labeling and Detection

To visualize antigen-antibody interactions, antibodies are often labeled with enzymes, fluorescent dyes, or radioactive isotopes. These labels generate a detectable signal when the antibody binds to the antigen, allowing researchers to measure the presence or concentration of the target molecule. Detection methods vary based on the type of immunochemical technique used, with colorimetric, fluorescent, and chemiluminescent detection being among the most common.

## 7.2 TYPES OF IMMUNOCHEMICAL TECHNIQUES

### 7.2.1 Enzyme-Linked Immunosorbent Assay (ELISA)

ELISA is one of the most popular immunochemical techniques due to its simplicity, sensitivity, and adaptability for high-throughput screening. It is widely used to detect and quantify antigens or antibodies in various samples, including blood, serum, and cell lysates. In ELISA, an antibody is immobilized on a solid surface, such as a microplate well. When the sample is added, the target antigen binds to the antibody. After washing away unbound molecules, a secondary antibody labeled with an enzyme is added. This enzyme reacts with a substrate to produce a detectable signal, such as a color change.

There are several types of ELISA, including direct, indirect, sandwich, and competitive ELISA, each suitable for different applications. For example, sandwich ELISA is ideal for detecting large proteins, while competitive ELISA is used to measure small molecules, such as hormones or drugs.

### 7.2.2 Western Blotting

Western blotting is a technique used primarily for detecting specific proteins within a sample. This technique combines protein separation by electrophoresis with antigen-antibody interactions. Proteins are first separated based on size using SDS-PAGE (sodium dodecyl sulfate-polyacrylamide gel electrophoresis). The separated proteins are then transferred to a membrane, where they are probed with specific antibodies. When the antibody binds to its target protein, it can be visualized using a secondary antibody conjugated to an enzyme or a fluorescent dye.

Western blotting is highly sensitive and provides information on protein size, abundance, and post-translational modifications. It is widely used in research and diagnostics, including for detecting viral infections, protein expression, and biomarker identification.

### 7.2.3 Immunohistochemistry (IHC)

Immunohistochemistry is used to detect and localize specific antigens within tissue sections. In IHC, tissue samples are treated with specific antibodies that bind to target proteins within the tissue. A secondary antibody conjugated to an enzyme or fluorescent dye is then added to generate a visible signal, allowing visualization of the antigen's distribution within the tissue. This method is widely used in pathology to diagnose diseases, including cancer, by identifying protein expression patterns in tissues. IHC provides spatial information on protein localization and expression within the cellular environment, making it a powerful tool in histopathology, neuroscience, and developmental biology.

### 7.2.4 Immunofluorescence

Immunofluorescence is similar to IHC but uses fluorescent dyes, or fluorophores, conjugated to antibodies to detect specific proteins. When exposed to certain wavelengths of light, these fluorophores emit fluorescence, which can be visualized using a fluorescence microscope. Immunofluorescence is commonly used to study protein localization in cells and tissues, enabling researchers to observe protein interactions and cellular structures. There are two types of immunofluorescence: direct, where the primary antibody is labeled with a fluorescent dye, and indirect, where a secondary antibody with a fluorescent tag binds to the primary antibody. Indirect immunofluorescence is often preferred for its higher sensitivity due to signal amplification.

### 7.2.5 Immunoprecipitation

Immunoprecipitation is a technique used to isolate and study specific proteins and their interacting partners. In this technique, an antibody that binds to a target protein is added to a sample, allowing the antibody-protein complex to form. The complex is then captured by protein A or G beads, which bind to the antibody's Fc region. After washing away unbound proteins, the isolated protein complex can be analyzed using techniques such as western blotting or mass spectrometry. Immunoprecipitation is widely used to study protein-protein interactions, post-translational modifications, and protein expression in cell and tissue lysates.

### 7.2.6 Flow Cytometry

Flow cytometry is a powerful immunochemical technique for analyzing the physical and chemical properties of cells in a fluid suspension. It uses fluorescently labeled antibodies to detect specific proteins on the surface or inside cells. As cells pass through a laser beam one by one, the fluorescence emitted by the labeled antibodies is measured, allowing researchers to analyze cell populations based on protein expression. Flow cytometry is widely used in immunology, hematology, and oncology to study cell types, immune responses, and disease progression. It is also a critical tool in clinical diagnostics for identifying immune cell subtypes and detecting abnormal cells.

## 7.3 APPLICATIONS OF IMMUNOCHEMICAL TECHNIQUES

Immunochemical techniques have a wide range of applications in research, diagnostics, and therapeutics, contributing significantly to advancements in medical science and biotechnology.

- **Disease Diagnosis**: Immunochemical techniques are essential in diagnosing infectious diseases, autoimmune disorders, and cancers. ELISA is commonly used for detecting viral infections, such as HIV and hepatitis, by identifying specific antibodies or antigens in patient serum. IHC and flow cytometry are valuable in diagnosing cancers, enabling pathologists to assess protein expression patterns in tumor cells and immune cell populations.
- **Drug Development**: Immunochemical techniques play a critical role in drug discovery and development. They are used to identify and validate

biomarkers, assess drug efficacy, and monitor protein expression in response to treatments. Western blotting and ELISA are frequently used to quantify drug targets and evaluate the effect of new compounds on protein expression.

- **Proteomics and Biomarker Discovery**: Immunochemical methods, such as western blotting and immunoprecipitation, are essential tools in proteomics research for identifying and characterizing proteins in complex biological samples. By enabling the analysis of protein expression, modification, and interaction, these techniques contribute to the discovery of disease-specific biomarkers and potential therapeutic targets.
- **Cell Biology and Immunology**: Immunofluorescence and flow cytometry are invaluable for studying cell structure, function, and signaling pathways. Immunofluorescence allows visualization of protein localization and cellular architecture, while flow cytometry enables quantitative analysis of immune cell populations, supporting research in immunology and cell biology.

## 7.4 ADVANTAGES AND LIMITATIONS OF IMMUNOCHEMICAL TECHNIQUES

Immunochemical techniques offer many advantages but also have some limitations.

### 7.4.1 Advantages

- **High Sensitivity and Specificity**: Immunochemical techniques provide high sensitivity and specificity, allowing the detection of low-abundance proteins.
- **Versatility**: These methods can be applied to various sample types, including blood, tissue, and cell lysates, making them widely applicable.
- **Quantitative and Qualitative Data**: Immunochemical techniques provide both quantitative and qualitative data on protein expression, interaction, and localization.

### 7.4.2 Limitations

- **Cross-Reactivity**: Antibodies may cross-react with non-target molecules, leading to false-positive results. Ensuring antibody specificity is crucial.
- **Time-Consuming**: Some immunochemical techniques, such as IHC and western blotting, can be labor-intensive and time-consuming.
- **Antibody Dependence**: The success of immunochemical techniques

depends on the quality and specificity of antibodies, which can be costly and difficult to obtain for some targets.

## 7.5 RECENT ADVANCES IN IMMUNOCHEMICAL TECHNIQUES

Recent advancements have improved the sensitivity, speed, and applicability of immunochemical techniques. High-throughput immunoassays, automated ELISA platforms, and multiplex assays allow simultaneous detection of multiple targets, increasing efficiency. Advanced imaging techniques, such as super-resolution microscopy, enhance the resolution of immunofluorescence, enabling detailed analysis of cellular structures. Additionally, developments in antibody engineering have led to the production of monoclonal antibodies with improved specificity and affinity.

## 7.6 CONCLUSION

Immunochemical techniques are integral to modern biological and medical research, providing critical insights into protein function, cellular pathways, and disease mechanisms. By leveraging the specificity of antigen-antibody interactions, these methods enable sensitive and accurate detection of biomolecules in complex biological samples. As technology advances, immunochemical techniques continue to evolve, offering new opportunities for research, diagnostics, and therapeutic applications. With their broad applicability and continued innovation, immunochemical methods will remain essential in scientific discovery and clinical practice.

# CHAPTER 8
# FLOW CYTOMETRY

Flow cytometry is a sophisticated and powerful analytical technique widely used in biomedical and biological research. It enables researchers to measure the physical and chemical characteristics of cells or particles suspended in a fluid as they pass through a laser beam.

The main advantage of flow cytometry is its ability to analyze large populations of cells individually, providing detailed quantitative data on parameters such as cell size, complexity, and specific molecular markers.

Flow cytometry has diverse applications in immunology, oncology, hematology, and microbiology, and is fundamental in both basic research and clinical diagnostics.

## 8.1 PRINCIPLES OF FLOW CYTOMETRY

Flow cytometry operates based on three main principles: fluidics, optics, and electronics.

- **Fluidics:** The fluidics system transports cells or particles in a fluid stream through the instrument. The cells pass through a narrow nozzle, ensuring they move in single file as they approach the laser beam. This arrangement allows the laser to illuminate one cell at a time, ensuring accurate detection and measurement of individual cells.
- **Optics:** The optics system consists of lasers, filters, and detectors. When a cell passes through the laser beam, it scatters light in different directions, and fluorescent molecules on the cell emit light at specific wavelengths. The scattered light and fluorescence signals are captured by detectors, which then transmit the information to the electronics system for analysis. Forward scatter (FSC) measures cell size, while side scatter (SSC) provides information on cell granularity or internal complexity.
- **Electronics:** The electronics system converts optical signals into digital data. The signals from the detectors are processed, quantified, and displayed as histograms or scatter plots. By analyzing these plots, researchers can assess characteristics like cell size, structure, and the presence of specific biomarkers based on fluorescence intensity.

## 8.2 FLUORESCENCE AND LABELING IN FLOW CYTOMETRY

One of the most critical aspects of flow cytometry is fluorescence, which enables the identification and quantification of specific proteins or markers on or inside cells. Fluorescent molecules, or fluorophores, are conjugated to antibodies that bind to target molecules. These labeled antibodies can then attach to specific cell surface or intracellular proteins, making it possible to detect and measure the abundance of these markers.

### 8.2.1 Fluorophores

A fluorophore (or fluorochrome, similarly to a chromophore) is a fluorescent chemical compound that can re-emit light upon light excitation. Fluorophores typically contain several combined aromatic groups, or planar or cyclic molecules with several $\pi$ bonds.

Fluorophores are sometimes used alone, as a tracer in fluids, as a dye for staining of certain structures, as a substrate of enzymes, or as a probe or indicator (when its fluorescence is affected by environmental aspects such as polarity or ions). More generally they are covalently bonded to macromolecules, serving as a markers (or dyes, or tags, or reporters) for affine or bioactive reagents (antibodies, peptides, nucleic acids). Fluorophores are notably used to stain tissues, cells, or materials in a variety of analytical methods, such as fluorescent imaging and spectroscopy.

Fluorescein, via its amine-reactive isothiocyanate derivative fluorescein isothiocyanate (FITC), has been one of the most popular fluorophores. From antibody labeling, the applications have spread to nucleic acids thanks to carboxyfluorescein. Other historically common fluorophores are derivatives of rhodamine (TRITC), coumarin, and cyanine. Newer generations of fluorophores, many of which are proprietary, often perform better, being more photostable, brighter, or less pH-sensitive than traditional dyes with comparable excitation and emission.

#### Fluorescence

The fluorophore absorbs light energy of a specific wavelength and re-emits light at a longer wavelength. The absorbed wavelengths, energy transfer efficiency, and time before emission depend on both the fluorophore structure and its chemical environment, since the molecule in its excited state interacts with surrounding molecules. Wavelengths of maximum absorption (≈ excitation) and emission (for example, Absorption/Emission = 485 nm/517 nm) are the typical terms used to refer to a given fluorophore, but the whole spectrum may be important to consider. The excitation wavelength spectrum may be a very narrow or broader band, or it may be all beyond a cutoff level. The emission spectrum is usually sharper than the excitation spectrum, and it is of a longer wavelength and correspondingly lower energy. Excitation energies range from ultraviolet through the visible spectrum, and emission energies may continue from visible light into the near infrared region.

The main characteristics of fluorophores are:

- **Maximum Excitation and Emission Wavelength (Expressed in Nanometers (nm)):** Corresponds to the peak in the excitation and emission spectra (usually one peak each).
- **Molar Absorption Coefficient (in $mol^{-1}cm^{-1}$):** Links the quantity of absorbed light, at a given wavelength, to the concentration of fluorophore in solution.
- **Quantum Yield:** Efficiency of the energy transferred from incident light to emitted fluorescence (the number of emitted photons per absorbed photons).
- **Lifetime (in Picoseconds):** Duration of the excited state of a fluorophore before returning to its ground state. It refers to the time taken for a population of excited fluorophores to decay to 1/e (≈0.368) of the original amount.
- **Stokes Shift:** The difference between the maximum excitation and maximum emission wavelengths.
- **Dark Fraction:** The proportion of the molecules not active in fluorescence emission. For quantum dots, prolonged single-molecule microscopy showed that 20-90% of all particles never emit fluorescence. On the other hand, conjugated polymer nanoparticles (Pdots) show almost no dark fraction in their fluorescence. Fluorescent proteins can have a dark fraction from protein misfolding or defective chromophore formation.

These characteristics drive other properties, including photobleaching or photoresistance (loss of fluorescence upon continuous light excitation). Other parameters should be considered, as the polarity of the fluorophore molecule, the fluorophore size and shape (i.e. for polarization fluorescence pattern), and other factors can change the behavior of fluorophores.

Fluorophores can also be used to quench the fluorescence of other fluorescent dyes or to relay their fluorescence at even longer wavelengths.

## Size (Molecular Weight)

Most fluorophores are organic small molecules of 20–100 atoms (200–1000 Dalton; the molecular weight may be higher depending on grafted modifications and conjugated molecules), but there are also much larger natural fluorophores that are proteins: green fluorescent protein (GFP) is 27 kDa, and several phycobiliproteins

(PE, APC...) are ≈240kDa. As of 2020, the smallest known fluorophore was claimed to be 3-hydroxyisonicotinaldehyde, a compound of 14 atoms and only 123 Da.

Fluorescence particles like quantum dots (2–10 nm diameter, 100–100,000 atoms) are also considered fluorophores.

The size of the fluorophore might sterically hinder the tagged molecule and affect the fluorescence polarity.

## Families

**Fig 1: Fluorescence of Different Substances under UV Light. Green is a Fluorescein, Red is Rhodamine B, Yellow is Rhodamine 6G, Blue is Quinine, Purple is a Mixture of Quinine and Rhodamine 6g. Solutions are about 0.001% Concentration in Water**

Fluorophore molecules could be either utilized alone, or serve as a fluorescent motif of a functional system. Based on molecular complexity and synthetic methods, fluorophore molecules could be generally classified into four categories: proteins and peptides, small organic compounds, synthetic oligomers and polymers, and multi-component systems.

Fluorescent proteins GFP, YFP, and RFP (green, yellow, and red, respectively) can be attached to other specific proteins to form a fusion protein, synthesized in cells after transfection of a suitable plasmid carrier.

Non-protein organic fluorophores belong to following major chemical families:

- **Xanthene Derivatives:** Fluorescein, rhodamine, Oregon green, eosin, and Texas red.
- **Cyanine Derivatives:** Cyanine, indocarbocyanine, oxacarbocyanine, thiacarbocyanine, and merocyanine.
- Squaraine derivatives and ring-substituted squaraines, including Seta and Square dyes.

- Squaraine rotaxane derivatives.
- Naphthalene derivatives (dansyl and prodan derivatives).
- Coumarin derivatives.
- **Oxadiazole Derivatives:** Pyridyloxazole, nitrobenzoxadiazole, and benzoxadiazole.
- **Anthracene Derivatives:** Anthraquinones, including DRAQ5, DRAQ7, and CyTRAK Orange.
- **Pyrene Derivatives:** Cascade blue, etc.
- **Oxazine Derivatives:** Nile red, Nile blue, cresyl violet, oxazine 170, etc.
- **Acridine Derivatives:** Proflavin, acridine orange, acridine yellow, etc.
- **Arylmethine Derivatives:** Auramine, crystal violet, malachite green.
- **Tetrapyrrole Derivatives:** Porphin, phthalocyanine, bilirubin.
- **Dipyrromethene Derivatives:** BODIPY, aza-BODIPY.

These fluorophores fluoresce due to delocalized electrons which can jump a band and stabilize the energy absorbed. For example, benzene, one of the simplest aromatic hydrocarbons, is excited at 254 nm and emits at 300 nm. This discriminates fluorophores from quantum dots, which are fluorescent semiconductor nanoparticles.

They can be attached to proteins to specific functional groups, such as amino groups (active ester, carboxylate, isothiocyanate, hydrazine), carboxyl groups (carbodiimide), thiol (maleimide, acetyl bromide), and organic azide (via click chemistry or non-specifically (glutaraldehyde)).

Additionally, various functional groups can be present to alter their properties, such as solubility, or confer special properties, such as boronic acid which binds to sugars or multiple carboxyl groups to bind to certain cations. When the dye contains an electron-donating and an electron-accepting group at opposite ends of the aromatic system, this dye will probably be sensitive to the environment's polarity (solvatochromic), hence called environment-sensitive. Often dyes are used inside cells, which are impermeable to charged molecules; as a result of this, the carboxyl groups are converted into an ester, which is removed by esterases inside the cells, e.g., fura-2AM and fluorescein-diacetate.

The following dye families are trademark groups, and do not necessarily share structural similarities.

- CF dye (Biotium).
- DRAQ and CyTRAK probes (BioStatus).

- BODIPY (Invitrogen).
- EverFluor (Setareh Biotech).
- Alexa Fluor (Invitrogen).
- Bella Fluor (Setareh Biotech).
- DyLight Fluor (Thermo Scientific, Pierce).
- Atto and Tracy (Sigma Aldrich).
- FluoProbes (Interchim).
- Abberior Dyes (Abberior).
- DY and MegaStokes Dyes (Dyomics).
- Sulfo Cy dyes (Cyandye).
- HiLyte Fluor (AnaSpec).
- Seta, SeTau and Square Dyes (SETA BioMedicals).
- Quasar and Cal Fluor dyes (Biosearch Technologies).
- SureLight Dyes (APC, RPEPerCP, Phycobilisomes) (Columbia Biosciences).
- APC, APCXL, RPE, BPE (Phyco-Biotech, Greensea, Prozyme, Flogen).
- Vio Dyes (Miltenyi Biotec).

## Applications

Fluorophores have particular importance in the field of biochemistry and protein studies, for example, in immunofluorescence, cell analysis, immunohistochemistry, and small molecule sensors.

## Uses Outside the Life Sciences

Fluorescent dyes find a wide use in industry, going under the name of "neon colors", such as:

- Multi-ton scale usages in textile dyeing and optical brighteners in laundry detergents.
- Advanced cosmetic formulations.
- Safety equipment and clothing.
- Organic light-emitting diodes (OLEDs).
- Fine arts and design (posters and paintings).
- Synergists for insecticides and experimental drugs.
- Dyes in highlighters to give off a glow-like effect.
- Solar panels to collect more light / wavelengths.

- Fluorescent sea dye is used to help airborne search and rescue teams locate objects in the water.

### 8.2.2 Multicolor Flow Cytometry

With advances in technology, multicolor flow cytometry allows the simultaneous detection of multiple fluorescent markers in a single sample. This enables detailed analysis of complex cellular populations, such as immune cells, by identifying subpopulations based on multiple surface or intracellular markers. This multiplexing capability is crucial for studying cellular diversity and functionality in immunology and cancer research.

## 8.3 KEY APPLICATIONS OF FLOW CYTOMETRY

Flow cytometry has a wide range of applications in both research and clinical diagnostics, with particular importance in immunology, oncology, and cell biology.

### 8.3.1 Immunology

Flow cytometry is a powerful technique used in immunology for the analysis of the physical and chemical characteristics of cells or particles suspended in a fluid. It allows for the rapid quantification and characterization of different cell populations based on their surface markers and other properties.

Flow cytometry has become an indispensable tool in immunology, providing critical insights into the cellular components of the immune system. Its ability to perform multiparametric analysis rapidly and accurately makes it valuable in both research and clinical applications, enhancing our understanding of immune functions and diseases.

As technology advances, the capabilities of flow cytometry continue to expand, further solidifying its role in the field of immunology.

#### Principles of Flow Cytometry

- **Fluidics System:** Cells are passed through a narrow stream in a fluid, allowing them to be analyzed one at a time.
- **Optics:** As cells flow through the laser beam, they scatter light. Forward scatter (FSC) correlates with cell size, while side scatter (SSC) correlates with cell granularity or internal complexity.
- **Fluorescence Detection:** Cells are typically stained with fluorescent antibodies that bind to specific cell surface or intracellular markers. When these cells pass through the laser, the fluorescent tags emit light at specific wavelengths, which is detected and analyzed.
- **Data Analysis:** The emitted light is collected by photomultiplier tubes and converted into digital signals. These signals can be analyzed to

determine the number of cells expressing particular markers, their relative fluorescence intensity, and various other parameters.

## Applications in Immunology

**Cell Population Identification:** Flow cytometry can distinguish between different cell types in a mixed population (e.g., lymphocytes, monocytes) based on surface markers like CD4, CD8, or CD19.

- **Cell Function Studies:** It can assess functional characteristics of immune cells, such as cytokine production, proliferation, and apoptosis, by using specific assays.
- **Immunophenotyping:** Identifying and quantifying the types of immune cells in blood or tissue samples, which is crucial in diagnosing and monitoring diseases like leukemia and lymphoma.
- **Clinical Applications:** Flow cytometry is used in clinical settings for blood and bone marrow analysis, monitoring of immune responses in patients (e.g., HIV, autoimmune diseases), and in hematopoietic stem cell transplantation.
- **Research Applications:** Researchers utilize flow cytometry to study immune responses, vaccine efficacy, and the immunological effects of drugs.
- **Single-Cell Analysis:** Flow cytometry allows for the examination of individual cells, which is essential for understanding heterogeneity in immune responses.

## Advantages and Limitations

### Advantages

- **Speed:** Can analyze thousands of cells per second.
- **Multiparametric Analysis:** Simultaneous detection of multiple markers on the same cell.
- **Quantitative Data:** Provides quantitative information about cell populations.

### Limitations

- **Complexity:** Requires specialized training and expertise to operate and analyze data.
- **Sample Preparation:** Cells must be adequately prepared and stained, which can introduce variability.
- **Limitations in Detection:** Certain low-abundance markers may be challenging to detect.

## 8.3.2 Oncology

Flow cytometry is an essential technology in oncology for the diagnosis, treatment monitoring, and research of cancer. It provides rapid and quantitative analysis of cellular characteristics, allowing for the detailed examination of tumor cells and their microenvironment.

Flow cytometry plays a crucial role in oncology, facilitating the identification and characterization of cancer cells, monitoring disease progression, and evaluating treatment responses. Its ability to provide rapid, quantitative, and multiparametric analysis makes it a vital tool in both clinical and research settings. As advances in flow cytometry technology continue, including the development of new fluorescent markers and improved analytical software, its applications in oncology are expected to expand, further enhancing cancer diagnosis, treatment, and research.

### Principles of Flow Cytometry

- **Fluidics System:** In flow cytometry, cells are suspended in a fluid stream and directed through a laser beam. This setup allows for the analysis of individual cells as they pass through the detection area.
- **Optical Detection:** When cells intersect with the laser light, they scatter light and may also emit fluorescence if they are labeled with fluorescent markers. The intensity of light scattering (FSC for size and SSC for granularity) provides information about cell morphology.
- **Fluorescent Labeling:** Tumor cells can be labeled with fluorescent antibodies that target specific tumor-associated antigens or surface markers, enabling identification of different cell populations and states.
- **Data Acquisition and Analysis:** The emitted fluorescence is collected and processed, allowing for the generation of histograms and dot plots that provide insight into the cellular composition of the sample.

### Applications in Oncology

- **Tumor Cell Identification:** Flow cytometry helps in identifying and characterizing tumor cells based on specific surface markers (e.g., CD markers, hormone receptors) that distinguish malignant cells from normal cells.
- **Immunophenotyping:** It is used to determine the immunophenotype of hematologic malignancies, such as leukemia and lymphoma. This characterization helps in classifying the type of cancer and informing treatment strategies.
- **Monitoring Minimal Residual Disease (MRD):** Flow cytometry is invaluable for detecting small numbers of residual cancer cells after

treatment, which can indicate relapse. This is particularly significant in leukemia management.

- **Prognostic and Predictive Biomarkers:** Flow cytometry can assess the expression of specific markers that provide prognostic information regarding disease outcomes and responses to therapy (e.g., P53, Ki-67).
- **Cell Cycle Analysis:** The technique allows for the analysis of cell cycle phases (G1, S, G2/M) in tumor cells, helping to assess the proliferation status of tumors.
- **Drug Response Assessment:** By using fluorescent probes, flow cytometry can evaluate the sensitivity of cancer cells to various chemotherapeutic agents, providing insights for personalized treatment approaches.
- **Tumor Microenvironment Studies:** Flow cytometry can analyze the immune cell populations within tumors, providing valuable information about the tumor microenvironment and immune evasion mechanisms.
- **Single-Cell Analysis:** Flow cytometry allows for the analysis of individual tumor cells, which is critical for understanding tumor heterogeneity and dynamics.

## Advantages and Limitations

### Advantages

- **High Throughput:** Capable of analyzing thousands of cells per second, making it efficient for large sample sizes.
- **Multiparametric Capabilities:** Enables simultaneous analysis of multiple markers, providing comprehensive cellular profiles.
- **Quantitative Data:** Offers quantitative assessments of cell populations, which is essential for clinical and research applications.

### Limitations

- **Sample Preparation Complexity:** Requires careful preparation and staining of samples, which can introduce variability.
- **High Cost:** The equipment and maintenance can be expensive, limiting access in some settings.
- **Data Interpretation Challenges:** Requires expertise to analyze and interpret complex data accurately.

## 8.3.3 Hematology

Flow cytometry is a pivotal technology in hematology, providing detailed analysis of blood cells and their properties. It enables the rapid and accurate assessment of cell populations, helping diagnose and monitor various hematological disorders, including anemias, leukemias, and lymphomas.

Flow cytometry has revolutionized hematology by enabling precise and rapid analysis of blood cells, which is critical for diagnosing and managing hematological diseases. Its ability to provide detailed insights into cell populations and functions makes it an invaluable tool in both clinical practice and research.

As advancements in technology continue, flow cytometry will likely play an even more significant role in the evolving field of hematology, improving patient outcomes and enhancing our understanding of blood disorders.

## Principles of Flow Cytometry

- **Fluidics System**: Blood cells are suspended in a fluid medium and directed through a narrow stream, allowing cells to pass one at a time through a laser beam.
- **Optical Detection**: When cells intersect with the laser, they scatter light. The forward scatter (FSC) correlates with cell size, while side scatter (SSC) provides information about cell granularity and internal complexity.
- **Fluorescent Labeling**: Cells can be stained with fluorescently labeled antibodies that bind to specific surface markers (such as CD markers), enabling the identification of different cell types.
- **Data Acquisition and Analysis:** The emitted fluorescence is collected by detectors and transformed into digital data, allowing for the generation of histograms and dot plots that illustrate the characteristics of the cell populations.

## Applications in Hematology

- **Cell Population Enumeration:** Flow cytometry allows for the precise counting of different types of blood cells (e.g., red blood cells, white blood cells, and platelets), which is essential for routine blood tests and diagnostics.
- **Immunophenotyping**: It is extensively used to classify hematologic malignancies (like leukemia and lymphoma) by analyzing the expression of surface markers. This classification helps in determining the type of cancer and guiding treatment.
- **Diagnosis of Hematological Disorders**: Flow cytometry can identify abnormalities in blood cell populations that are characteristic of various disorders, including:

- *Leukemias:* Identifying specific leukemia types based on cell surface marker expression.
- *Lymphomas:* Characterizing lymphoid neoplasms and distinguishing between different subtypes.
- *Anemias:* Assessing red blood cell morphology and identifying specific types of anemia (e.g., hemolytic anemia).

- **Assessment of Minimal Residual Disease (MRD):** Flow cytometry is critical for detecting small numbers of residual cancer cells after treatment, particularly in acute lymphoblastic leukemia (ALL) and acute myeloid leukemia (AML). MRD assessment is vital for predicting relapse.
- **Platelet Function Analysis:** Flow cytometry can evaluate platelet activation and function, which is crucial for understanding bleeding disorders and thrombotic conditions.
- **Stem Cell Analysis:** In hematopoietic stem cell transplantation, flow cytometry is used to analyze stem cell products and monitor engraftment by assessing the recovery of blood cell populations.
- **Single-Cell Analysis:** Flow cytometry enables the analysis of individual blood cells, providing insights into heterogeneity within populations, which is particularly useful in research.

### Advantages and Limitations

#### Advantages

- **Speed:** Capable of analyzing thousands of cells per second, facilitating rapid diagnosis and monitoring.
- **Multiparametric Analysis:** Simultaneously assesses multiple markers, providing comprehensive information about cell populations.
- **Quantitative Data:** Delivers quantitative assessments, enhancing the accuracy of results.

#### Limitations

- **Complexity:** Requires specialized training for both operation and data interpretation.
- **Sample Preparation:** Proper staining and handling of samples are crucial to avoid variability in results.
- **Cost:** Flow cytometry equipment and reagents can be expensive, which may limit access in some laboratories.

### 8.3.4 Microbiology

Flow cytometry is a versatile tool in microbiology, allowing researchers and clinicians to analyze microbial populations, assess their characteristics, and study interactions among different microbial species. The ability to examine cells at high speed and in a multiparametric manner makes flow cytometry particularly valuable in both research and clinical microbiology. Flow cytometry is a powerful tool in microbiology, enabling the rapid and detailed analysis of microbial populations and their characteristics. Its applications range from clinical diagnostics to environmental monitoring, making it invaluable for researchers and clinicians alike. As advancements in flow cytometry technology continue, including improvements in fluorescent markers and data analysis techniques, its applications in microbiology are expected to expand, enhancing our understanding of microbial dynamics and interactions in various settings.

## Principles of Flow Cytometry

- **Fluidics System:** In flow cytometry, samples containing microbial cells (bacteria, yeast, or other microorganisms) are suspended in a fluid and passed through a laser beam, which enables single-cell analysis.
- **Optical Detection:** As the cells flow past the laser, they scatter light. The forward scatter (FSC) provides information about cell size, while side scatter (SSC) relates to cell granularity or internal complexity.
- **Fluorescent Labeling:** Microbial cells can be labeled with fluorescent dyes or antibodies that bind to specific cell surface markers or intracellular components. This allows for the identification and characterization of different microbial species or states (e.g., viable vs. non-viable).
- **Data Acquisition and Analysis:** The fluorescence emitted from labeled cells is detected and analyzed, providing detailed information on the cell populations present in the sample.

## Applications in Microbiology

- **Microbial Identification and Classification:** Flow cytometry can distinguish between different microbial species based on their unique surface markers. This application is particularly useful in identifying pathogens in clinical samples or environmental samples.
- **Viability and Cell Health Assessment:** By using specific fluorescent dyes, flow cytometry can determine cell viability, assessing whether cells are alive, dead, or metabolically active. Common dyes include propidium iodide (PI) and 7-AAD, which can differentiate live and dead cells.
- **Bacterial Cell Counting:** Flow cytometry provides rapid and accurate

enumeration of bacterial cells, which is essential for various applications, including water quality testing, food safety, and clinical diagnostics.

- **Antimicrobial Susceptibility Testing**: Flow cytometry can be used to assess the effects of antibiotics on bacterial populations, allowing for rapid determination of susceptibility or resistance profiles.
- **Biofilm Analysis**: Flow cytometry can evaluate biofilm formation by measuring the properties of cells in biofilms, such as their size, density, and viability. This is crucial for understanding biofilm-related infections.
- **Single-Cell Analysis:** The technique allows researchers to examine individual microbial cells, providing insights into heterogeneity within populations, which is important for understanding microbial ecology and evolution.
- **Detection of Intracellular Markers**: Flow cytometry can be used to study intracellular processes, such as the production of specific metabolites or the presence of signaling molecules, helping researchers understand microbial physiology.
- **Immune Response Studies**: In the context of infectious diseases, flow cytometry can analyze immune responses to microbial infections by measuring cytokine production, T cell activation, and other immune parameters.
- **Environmental Microbiology**: Flow cytometry is useful for analyzing microbial communities in environmental samples, such as soil or water, allowing for the assessment of biodiversity and the impact of environmental changes.

## Advantages and Limitations

### Advantages

- **High Throughput**: Capable of analyzing thousands of cells per second, making it efficient for large sample sizes.
- **Multiparametric Capability**: Simultaneously analyzes multiple characteristics of individual cells, providing comprehensive data.
- **Quantitative Analysis:** Delivers quantitative measurements of cell populations, enhancing the reliability of results.

### Limitations

- **Sample Preparation**: Requires careful preparation and staining, which can introduce variability if not done correctly.

- **Technical Expertise:** Requires trained personnel for operation and data interpretation.
- **Cost:** The cost of flow cytometry equipment and reagents can be significant, potentially limiting accessibility.

## 8.4 DATA ANALYSIS AND INTERPRETATION IN FLOW CYTOMETRY

Data analysis in flow cytometry involves interpreting histograms and scatter plots, where data points represent individual cells.

Key parameters include forward scatter (FSC), side scatter (SSC), and fluorescence intensity.

- **Gating:** Gating is a critical step in data analysis, where specific cell populations are selected based on defined criteria. For example, researchers can create gates to isolate lymphocytes from other blood cells, based on their unique FSC and SSC characteristics. Gating enables focused analysis on the population of interest, improving data accuracy.
- **Scatter Plots and Histograms:** Scatter plots show two parameters simultaneously, such as FSC vs. SSC, allowing researchers to distinguish between cell types. Histograms display the distribution of a single parameter, such as fluorescence intensity, which is useful for quantifying marker expression on cell surfaces or within cells.
- **Flow Cytometry Software:** Specialized software, such as FlowJo and FCS Express, assists in data visualization and statistical analysis. These programs allow for complex gating strategies, multicolor analysis, and detailed reporting, making data interpretation more efficient and reproducible.

## 8.5 ADVANTAGES AND LIMITATIONS OF FLOW CYTOMETRY

Flow cytometry offers numerous advantages but also has limitations.

### 8.5.1 Advantages

- **High Sensitivity and Specificity:** Flow cytometry can detect rare cell populations and analyze multiple markers simultaneously.
- **Speed and Efficiency:** Flow cytometry can process thousands of cells per second, providing rapid results.
- **Quantitative and Multiparametric Analysis:** Flow cytometry provides quantitative data on multiple parameters, enabling comprehensive analysis of cell characteristics.

### 8.5.2 Limitations

- **Complexity and Cost:** Flow cytometers are expensive and require skilled personnel for operation and data analysis.
- **Sample Preparation:** Flow cytometry requires well-prepared single-cell suspensions, which can be challenging for certain tissue types.
- **Potential for Artifacts:** Non-specific binding and autofluorescence can interfere with data accuracy, especially in multicolor experiments.

## 8.6 ADVANCES IN FLOW CYTOMETRY TECHNOLOGY

Recent advancements in flow cytometry have significantly improved its capabilities, including the development of spectral flow cytometry and mass cytometry.

- **Spectral Flow Cytometry:** Traditional flow cytometers detect specific wavelengths of fluorescence, but spectral flow cytometers analyze the full emission spectrum of each fluorophore. This allows for the use of more fluorophores in a single experiment, enhancing multiplexing capabilities and enabling more detailed analyses of complex cell populations.
- **Mass Cytometry (CyTOF):** Mass cytometry combines flow cytometry with mass spectrometry to overcome the limitations of fluorescence overlap. In mass cytometry, antibodies are tagged with heavy metal isotopes instead of fluorophores. This technique allows for the simultaneous analysis of over 40 markers without fluorescence interference, making it a powerful tool for studying complex cellular systems in immunology and oncology.

## 8.7 FUTURE PROSPECTS AND APPLICATIONS OF FLOW CYTOMETRY

Flow cytometry continues to evolve, with new technologies enhancing its versatility and application range.

Advances in automation, machine learning, and artificial intelligence are expected to improve data analysis and enable high-throughput screening, making flow cytometry more accessible and efficient. Future applications may include single-cell RNA sequencing integration, enhanced imaging capabilities, and personalized medicine, where flow cytometry will play a role in tailoring treatments based on individual cellular profiles.

## 8.8 CONCLUSION

Flow cytometry is a versatile and powerful technology that has become indispensable in biological and medical research.

By providing rapid, high-throughput analysis of individual cells, it allows researchers to explore cellular diversity and gain insights into complex biological processes.

Flow cytometry's applications in immunology, oncology, hematology, and microbiology highlight its critical role in advancing scientific understanding and improving clinical diagnostics. As technology advances, flow cytometry is likely to become even more integral in research and healthcare, supporting innovations in fields like immunotherapy, cancer treatment, and personalized medicine.

# CHAPTER 9

# RADIOISOTOPE TECHNIQUES

Radioisotope techniques have been widely adopted in various fields, especially in medical and biological research, for their ability to trace, measure, and analyze chemical processes at the molecular level. These techniques involve the use of radioactive isotopes as tracers, enabling scientists to study biological, chemical, and physical processes with high specificity and sensitivity.

By incorporating radioisotopes into molecules of interest, researchers can observe and quantify interactions, metabolic pathways, and cellular processes with remarkable precision. The application of radioisotope techniques has expanded into areas such as cancer diagnosis and treatment, pharmacology, and environmental studies, illustrating its broad utility in both scientific research and clinical applications.

## 9.1 PRINCIPLES OF RADIOISOTOPE TECHNIQUES

Radioisotope techniques are based on the principle of using unstable isotopes, or radioisotopes, that emit radiation as they decay into a stable form. Common types of radiation include alpha particles, beta particles, and gamma rays, which vary in their energy, range, and penetrative ability.

### 9.1.1 Alpha Particles

Alpha particles, also called alpha rays or alpha radiation, consist of two protons and two neutrons bound together into a particle identical to a helium-4 nucleus. They are generally produced in the process of alpha decay but may also be produced in other ways. Alpha particles are named after the first letter in the Greek alphabet, $\alpha$. The symbol for the alpha particle is $\alpha$ or $\alpha^{2+}$. Because they are identical to helium nuclei, they are also sometimes written as $He^{2+}$ or ${}^{4}_{2}He^{2+}$ indicating a helium ion with a +2 charge (missing its two electrons). Once the ion gains electrons from its environment, the alpha particle becomes a normal (electrically neutral) helium atom ${}^{4}_{2}He$.

Alpha particles have a net spin of zero. When produced in standard alpha radioactive decay, alpha particles generally have a kinetic energy of about 5 MeV and a velocity in the vicinity of 4% of the speed of light. They are a highly ionizing form of particle radiation, with low penetration depth (stopped by a few centimetres of air, or by the skin).

However, so-called long-range alpha particles from ternary fission are three times as energetic and penetrate three times as far. The helium nuclei that form 10–12% of cosmic rays are also usually of much higher energy than those produced by nuclear decay processes, and thus may be highly penetrating and able to traverse the human body and also many metres of dense solid shielding, depending on their energy. To a lesser extent, this is also true of very high-energy helium nuclei produced by particle accelerators.

## Alpha Decay

The best-known source of alpha particles is alpha decay of heavier (mass number of at least 104) atoms. When an atom emits an alpha particle in alpha decay, the atom's mass number decreases by four due to the loss of the four nucleons in the alpha particle. The atomic number of the atom goes down by two, as a result of the loss of two protons – the atom becomes a new element. Examples of this sort of nuclear transmutation by alpha decay are the decay of uranium to thorium, and that of radium to radon.

Alpha particles are commonly emitted by all of the larger radioactive nuclei such as uranium, thorium, actinium, and radium, as well as the transuranic elements. Unlike other types of decay, alpha decay as a process must have a minimum-size atomic nucleus that can support it. The smallest nuclei that have to date been found to be capable of alpha emission are beryllium-8 and tellurium-104, not counting beta-delayed alpha emission of some lighter elements. The alpha decay sometimes leaves the parent nucleus in an excited state; the emission of a gamma ray then removes the excess energy.

### Mechanism of Production in Alpha Decay

In contrast to beta decay, the fundamental interactions responsible for alpha decay are a balance between the electromagnetic force and nuclear force. Alpha decay results from the Coulomb repulsion between the alpha particle and the rest of the nucleus, which both have a positive electric charge, but which is kept in check by the nuclear force. In classical physics, alpha particles do not have enough energy to escape the potential well from the strong force inside the nucleus (this well involves escaping the strong force to go up one side of the well, which is followed by the electromagnetic force causing a repulsive push-off down the other side).

However, the quantum tunnelling effect allows alphas to escape even though they do not have enough energy to overcome the nuclear force. This is allowed by the wave nature of matter, which allows the alpha particle to spend some of its time in a region so far from the nucleus that the potential from the repulsive electromagnetic force has fully compensated for the attraction of the nuclear force. From this point, alpha particles can escape.

## Ternary Fission

Especially energetic alpha particles deriving from a nuclear process are produced in the relatively rare (one in a few hundred) nuclear fission process of ternary fission. In this process, three charged particles are produced from the event instead of the normal two, with the smallest of the charged particles most probably (90% probability) being an alpha particle. Such alpha particles are termed "long range alphas" since at their typical energy of 16 MeV, they are at far higher energy than is ever produced by alpha decay. Ternary fission happens in both neutron-induced fission (the nuclear reaction that happens in a nuclear reactor), and also when fissionable and fissile actinides nuclides (i.e., heavy atoms capable of fission) undergo spontaneous fission as a form of radioactive decay. In both induced and spontaneous fission, the higher energies available in heavy nuclei result in long range alphas of higher energy than those from alpha decay.

## Accelerators

Energetic helium nuclei (helium ions) may be produced by cyclotrons, synchrotrons, and other particle accelerators. Convention is that they are not normally referred to as "alpha particles".

## Solar Core Reactions

Helium nuclei may participate in nuclear reactions in stars, and occasionally and historically these have been referred to as alpha reactions.

## Cosmic Rays

In addition, extremely high energy helium nuclei sometimes referred to as alpha particles make up about 10 to 12% of cosmic rays. The mechanisms of cosmic ray production continue to be debated.

## Energy and Absorption

The energy of the alpha particle emitted in alpha decay is mildly dependent on the half-life for the emission process, with many orders of magnitude differences in half-life being associated with energy changes of less than 50%, shown by the Geiger–Nuttall law.

The energy of alpha particles emitted varies, with higher energy alpha particles being emitted from larger nuclei, but most alpha particles have energies of between 3 and 7MeV (mega-electron-volts), corresponding to extremely long and extremely short half-lives of alpha-emitting nuclides, respectively. The energies and ratios are often distinct and can be used to identify specific nuclides as in alpha spectrometry. With a typical kinetic energy of 5 MeV; the speed of emitted alpha particles is 15,000 km/s, which is 5% of the speed of light. This energy is a substantial amount of energy for a single particle, but their high mass means alpha particles have a lower speed than any other common type of radiation, e.g. β particles, neutrons.

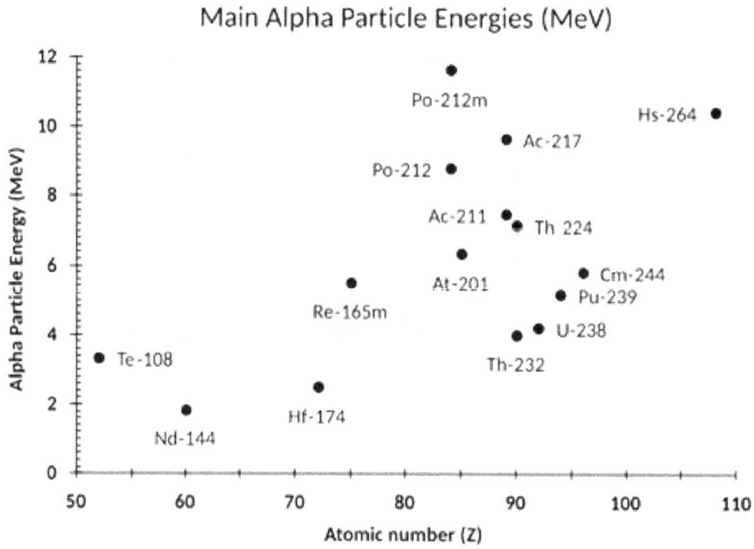


**Fig 1 : Example Selection of Radioactive Nuclides with Main Emitted Alpha Particle Energies Plotted against their Atomic Number. Each Nuclide has a Distinct Alpha Spectrum**

Because of their charge and large mass, alpha particles are easily absorbed by materials, and they can travel only a few centimetres in air. They can be absorbed by tissue paper or by the outer layers of human skin. They typically penetrate skin about 40 micrometres, equivalent to a few cells deep.

## Biological Effects

Due to the short range of absorption and inability to penetrate the outer layers of skin, alpha particles are not, in general, dangerous to life unless the source is ingested or inhaled. Because of this high mass and strong absorption, if alpha-emitting radionuclides do enter the body (upon being inhaled, ingested, or injected, as with the use of Thorotrast for high-quality X-ray images prior to the 1950s), alpha radiation is the most destructive form of ionizing radiation.

It is the most strongly ionizing, and with large enough doses can cause any or all of the symptoms of radiation poisoning. It is estimated that chromosome damage from alpha particles is anywhere from 10 to 1000 times greater than that caused by an equivalent amount of gamma or beta radiation, with the average being set at 20 times. A study of European nuclear workers exposed internally to alpha radiation from plutonium and uranium found that when relative biological effectiveness is considered to be 20, the carcinogenic potential (in terms of lung cancer) of alpha radiation appears to be consistent with that reported for doses of external gamma radiation i.e. a given dose of alpha-particles inhaled presents the same risk as a 20-times higher dose of gamma radiation.

The powerful alpha emitter polonium-210 (a milligram of $^{210}Po$ emits as many alpha particles per second as 4.215 grams of $^{226}Ra$) is suspected of playing a role in lung cancer and bladder cancer related to tobacco smoking. $^{210}Po$ was used to kill Russian dissident and ex-FSB officer Alexander V. Litvinenko in 2006.

## History of Discovery and Use

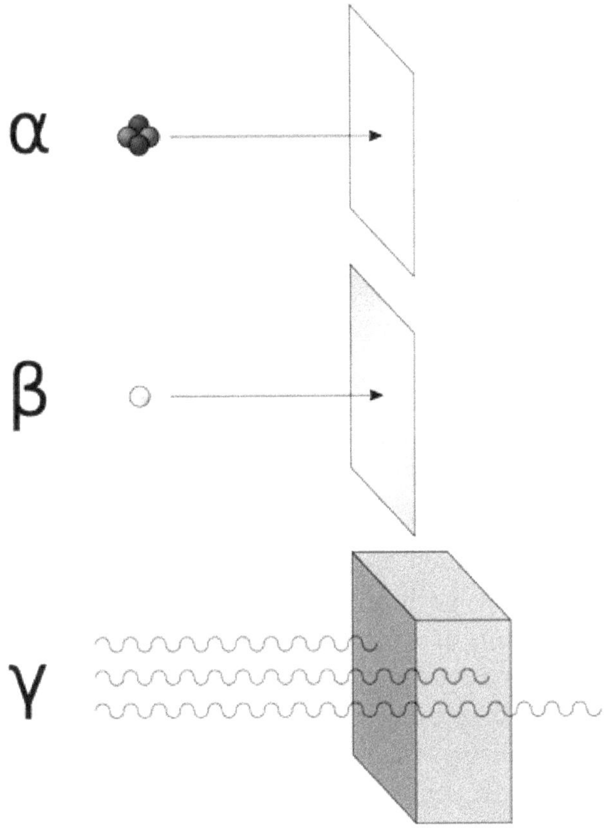


**Fig 2: Alpha Radiation Consists of Helium-4 Nucleus and is Readily Stopped by a Sheet of Paper. Beta Radiation, Consisting of Electrons, is Halted by an Aluminum Plate. Gamma Radiation is Eventually Absorbed as it Penetrates a Dense Material. Lead is Good at Absorbing Gamma Radiation, Due to its Density**

In 1899, physicists Ernest Rutherford (working in McGill University in Montreal, Canada) and Paul Villard (working in Paris) separated radiation into three types: eventually named alpha, beta, and gamma by Rutherford, based on penetration of objects and deflection by a magnetic field. Alpha rays were defined by Rutherford as those having the lowest penetration of ordinary objects.

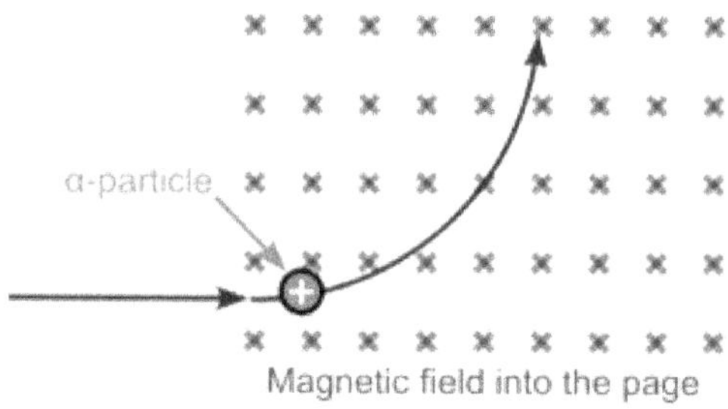


**Fig 3: An Alpha Particle is Deflected by a Magnetic Field**

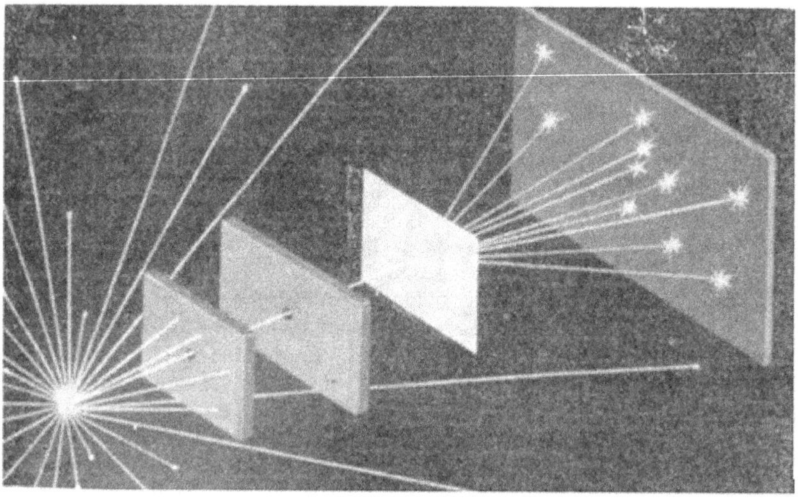

**Fig 4: Dispersing of Alpha Particles on a Thin Metal Sheet**

Rutherford's work also included measurements of the ratio of an alpha particle's mass to its charge, which led him to the hypothesis that alpha particles were doubly charged helium ions (later shown to be bare helium nuclei). In 1909, Ernest Rutherford and Thomas Royds finally proved that alpha particles were indeed helium ions. To do this they collected and purified the gas emitted by radium, a known alpha particle emitter, in a glass tube. An electric spark discharge inside the tube produced light. Subsequent study of the spectra of this light showed that the gas was helium and thus the alpha particles were indeed the helium ions.

Because alpha particles occur naturally, but can have energy high enough to participate in a nuclear reaction, study of them led to much early knowledge of nuclear physics. Rutherford used alpha particles emitted by radium bromide to infer that J. J. Thomson's Plum pudding model of the atom was fundamentally flawed. In Rutherford's gold foil experiment conducted by his students Hans Geiger and Ernest Marsden, a narrow beam of alpha particles was established, passing through very thin (a few hundred atoms thick) gold foil. The alpha particles were detected by a zinc sulfide screen, which emits a flash of light upon an alpha particle collision. Rutherford hypothesized that, assuming the "plum pudding" model of the atom was correct, the positively charged alpha particles would be only slightly deflected, if at all, by the dispersed positive charge predicted.

It was found that some of the alpha particles were deflected at much larger angles than expected (at a suggestion by Rutherford to check it) and some even bounced almost directly back.

Although most of the alpha particles went straight through as expected, Rutherford commented that the few particles that were deflected was akin to shooting a fifteen-inch shell at tissue paper only to have it bounce off, again assuming the "plum pudding" theory was correct. It was determined that the atom's positive charge was concentrated in a small area in its center, making the positive charge dense enough to deflect any positively charged alpha particles that came close to what was later termed the nucleus.

Prior to this discovery, it was not known that alpha particles were themselves atomic nuclei, nor was the existence of protons or neutrons known. After this discovery, J.J. Thomson's "plum pudding" model was abandoned, and Rutherford's experiment led to the Bohr model and later the modern wave-mechanical model of the atom.

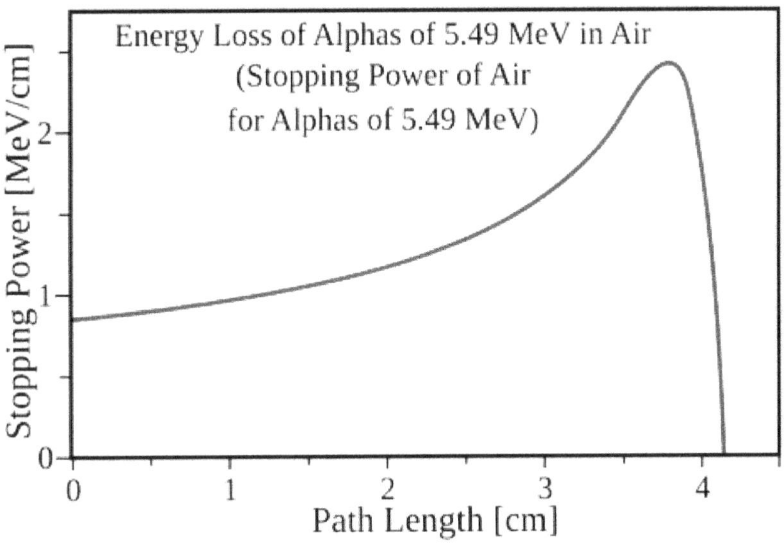


**Fig 5: Energy-Loss (Bragg Curve) in Air for Typical Alpha Particle Emitted through Radioactive Decay**

In 1917, Rutherford went on to use alpha particles to accidentally produce what he later understood as a directed nuclear transmutation of one element to another. Transmutation of elements from one to another had been understood since 1901 as a result of natural radioactive decay, but when Rutherford projected alpha particles from alpha decay into air, he discovered this produced a new type of radiation which proved to be hydrogen nuclei (Rutherford named these protons).

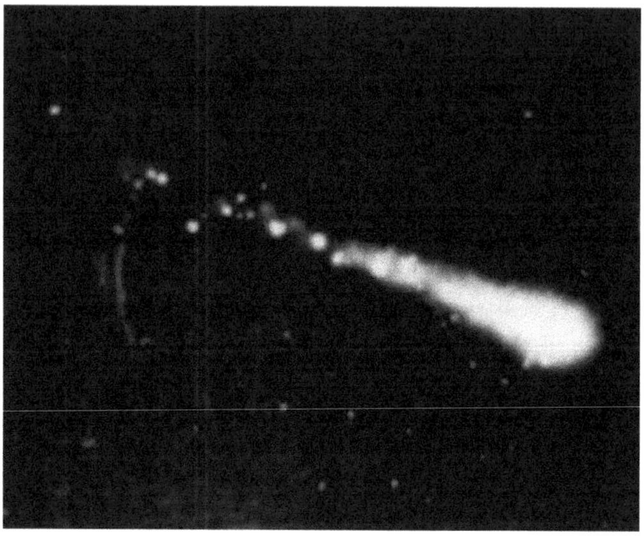

**Fig 6: The Trace of a Single Alpha Particle Obtained by Nuclear Physicist Wolfhart Willimczik with his Spark Chamber Specially Made for Alpha Particles**

Further experimentation showed the protons to be coming from the nitrogen component of air, and the reaction was deduced to be a transmutation of nitrogen into oxygen in the reaction

$$^{14}N + \alpha \rightarrow ^{17}O + p$$

This was the first discovered nuclear reaction. To the adjacent pictures: According to the energy-loss curve by Bragg, it is recognizable that the alpha particle indeed loses more energy on the end of the trace.

## Anti-Alpha Particle

In 2011, members of the international STAR collaboration using the Relativistic Heavy Ion Collider at the U.S. Department of Energy's Brookhaven National Laboratory detected the antimatter partner of the helium nucleus, also known as the anti-alpha. The experiment used gold ions moving at nearly the speed of light and colliding head on to produce the antiparticle.

## Applications

### Devices

- Some smoke detectors contain a small amount of the alpha emitter americium-241. The alpha particles ionize air within a small gap. A small current is passed through that ionized air. Smoke particles from fire that enter the air gap reduce the current flow, sounding the alarm. The isotope is extremely dangerous if inhaled or ingested, but the danger is minimal if the source is kept sealed. Many municipalities have established programs to collect and dispose of old smoke detectors, to keep them out of the general waste stream. However the US EPA says they "may be thrown away with household garbage".

- Alpha decay can provide a safe power source for radioisotope thermoelectric generators used for space probes. Alpha decay is much more easily shielded against than other forms of radioactive decay. Plutonium-238, a source of alpha particles, requires only 2.5 mm of lead shielding to protect against unwanted radiation.
- Static eliminators typically use polonium-210, an alpha emitter, to ionize air, allowing the "static cling" to more rapidly dissipate.

### Cancer Treatment

Alpha-emitting radionuclides are presently being used in three different ways to eradicate cancerous tumors: as an infusible radioactive treatment targeted to specific tissues (radium-223), as a source of radiation inserted directly into solid tumors (radium-224), and as an attachment to an tumor-targeting molecule, such as an antibody to a tumor-associated antigen.

Radium-223 is an alpha emitter that is naturally attracted to the bone because it is a calcium mimetic. Radium-223 (as radium-223 dichloride) can be infused into a cancer patient's veins, after which it migrates to parts of the bone where there is rapid turnover of cells due to the presence of metastasized tumors. Once within the bone, Ra-223 emits alpha radiation that can destroy tumor cells within a 100-micron distance.

This approach has been in use since 2013 to treat prostate cancer which has metastasized to the bone. Radionuclides infused into the circulation are able to reach sites that are accessible to blood vessels. This means, however, that the interior of a large tumor that is not vascularized (i.e. is not well penetrated by blood vessels) may not be effectively eradicated by the radioactivity.

Radium-224 is a radioactive atom that is utilized as a source of alpha radiation in a cancer treatment device called DaRT (diffusing alpha emitters radiation therapy). Each radium-224 atom undergoes a decay process producing 6 daughter atoms. During this process, 4 alpha particles are emitted. The range of an alpha particle—up to 100 microns—is insufficient to cover the width of many tumors. However, radium-224's daughter atoms can diffuse up to 2–3 mm in the tissue, thus creating a "kill region" with enough radiation to potentially destroy an entire tumor, if the seeds are placed appropriately. Radium-224's half-life is short enough at 3.6 days to produce a rapid clinical effect while avoiding the risk of radiation damage due to overexposure. At the same time, the half-life is long enough to allow for handling and shipping the seeds to a cancer treatment center at any location across the globe.

Targeted alpha therapy for solid tumors involves attaching an alpha-particle-emitting radionuclide to a tumor-targeting molecule such as an antibody, that can be delivered by intravenous administration to a cancer patient.

### Alpha Radiation and DRAM Errors

In computer technology, dynamic random access memory (DRAM) "soft errors" were linked to alpha particles in 1978 in Intel's DRAM chips.

The discovery led to strict control of radioactive elements in the packaging of semiconductor materials, and the problem is largely considered to be solved.

## 9.1.2 Beta Particles

A beta particle, also called beta ray or beta radiation (symbol β), is a high-energy, high-speed electron or positron emitted by the radioactive decay of an atomic nucleus, known as beta decay. There are two forms of beta decay, $\beta^-$ decay and $\beta^+$ decay, which produce electrons and positrons, respectively.

Beta particles with an energy of 0.5 MeV have a range of about one metre in the air; the distance is dependent on the particle's energy and the air's density and composition.

Beta particles are a type of ionizing radiation, and for radiation protection purposes, they are regarded as being more ionising than gamma rays, but less ionising than alpha particles. The higher the ionising effect, the greater the damage to living tissue, but also the lower the penetrating power of the radiation through matter.

### Beta Decay Modes

#### β- Decay (Electron Emission)

An unstable atomic nucleus with an excess of neutrons may undergo $\beta^-$ decay, where a neutron is converted into a proton, an electron, and an electron antineutrino (the antiparticle of the neutrino):

$$n \rightarrow p + e^- + \nu_e$$

This process is mediated by the weak interaction. The neutron turns into a proton through the emission of a virtual $W^-$ boson.

At the quark level, $W^-$ emission turns a down quark into an up quark, turning a neutron (one up quark and two down quarks) into a proton (two up quarks and one down quark). The virtual $W^-$ boson then decays into an electron and an antineutrino.

β- decay commonly occurs among the neutron-rich fission byproducts produced in nuclear reactors. Free neutrons also decay via this process. Both of these processes contribute to the copious quantities of beta rays and electron antineutrinos produced by fission-reactor fuel rods.

### β+ Decay (Positron Emission)

Unstable atomic nuclei with an excess of protons may undergo $\beta^+$ decay, also called positron decay, where a proton is converted into a neutron, a positron, and an electron neutrino:

$$p \rightarrow n + e^+ + \nu_e$$

Beta-plus decay can only happen inside nuclei when the absolute value of the binding energy of the daughter nucleus is greater than that of the parent nucleus, i.e., the daughter nucleus is a lower-energy state.

### Beta Decay Schemes

The accompanying decay scheme diagram shows the beta decay of caesium-137. $^{137}$Cs is noted for a characteristic gamma peak at 661 keV, but this is actually emitted by the daughter radionuclide $^{137m}$Ba. The diagram shows the type and energy of the emitted radiation, its relative abundance, and the daughter nuclides after decay.

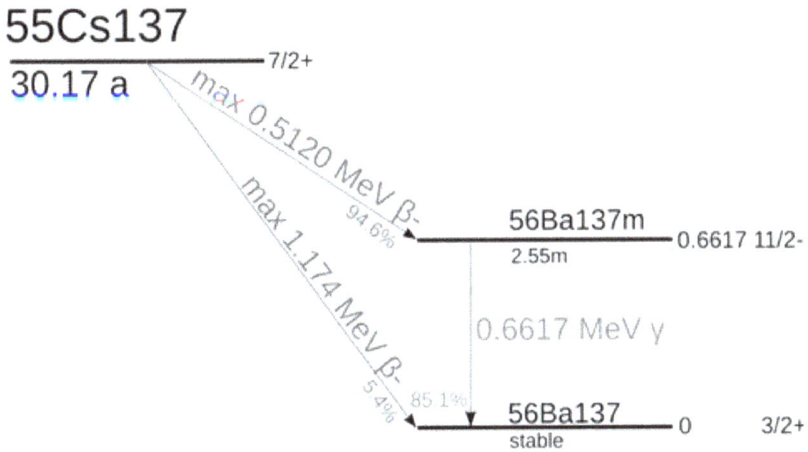


**Fig 7: Caesium-137 Decay Scheme, Showing it Initially Undergoes Beta Decay. The 661 keV Gamma Peak Associated with $^{137}$Cs is Actually Emitted by the Daughter Radionuclide**

Phosphorus-32 is a beta emitter widely used in medicine. It has a short half-life of 14.29 days and decays into sulfur-32 by beta decay as shown in this nuclear equation:

$$^{32}_{15}P \rightarrow {}^{32}_{16}S^{1+} + e^- + \nu_e$$

1.709 MeV of energy is released during the decay. The kinetic energy of the electron varies with an average of approximately 0.5 MeV and the remainder of the energy is carried by the nearly undetectable electron antineutrino.

In comparison to other beta radiation-emitting nuclides, the electron is moderately energetic. It is blocked by around 1 m of air or 5 mm of acrylic glass.

## Interaction with other Matter

Of the three common types of radiation given off by radioactive materials, alpha, beta and gamma, beta has the medium penetrating power and the medium ionising power. Although the beta particles given off by different radioactive materials vary in energy, most beta particles can be stopped by a few millimeters of aluminium. However, this does not mean that beta-emitting isotopes can be completely shielded by such thin shields: as they decelerate in matter, beta electrons emit secondary gamma rays, which are more penetrating than betas per se. Shielding composed of materials with lower atomic weight generates gammas with lower energy, making such shields somewhat more effective per unit mass than ones made of larger atoms such as lead.

Being composed of charged particles, beta radiation is more strongly ionizing than gamma radiation. When passing through matter, a beta particle is decelerated by electromagnetic interactions and may give off bremsstrahlung X-rays.

In water, beta radiation from many nuclear fission products typically exceeds the speed of light in that material (which is about 75% that of light in vacuum), and thus generates blue Cherenkov radiation when it passes through water. The intense beta radiation from the fuel rods of swimming pool reactors can thus be visualized through the transparent water that covers and shields the reactor.

### Beta Spectroscopy

The energy contained within individual beta particles is measured via *beta spectrometry*; the study of the obtained distribution of energies as a spectrum is *beta spectroscopy*. Determination of this energy is done by measuring the amount of deflection of the electron's path under a magnetic field.

## Applications

Beta particles can be used to treat health conditions such as eye and bone cancer and are also used as tracers. Strontium-90 is the material most commonly used to produce beta particles.

Beta particles are also used in quality control to test the thickness of an item, such as paper, coming through a system of rollers. Some of the beta radiation is absorbed while passing through the product. If the product is made too thick or thin, a correspondingly different amount of radiation will be absorbed. A computer program monitoring the quality of the manufactured paper will then move the rollers to change the thickness of the final product.

An illumination device called a *betalight* contains tritium and a phosphor. As tritium decays, it emits beta particles; these strike the phosphor, causing the phosphor to give off photons, much like the cathode-ray tube in a television. The illumination requires no external power, and will continue as long as the tritium exists (and the phosphors do not themselves chemically change); the amount of light produced will drop to half its original value in 12.32 years, the half-life of tritium. Beta-plus (or positron) decay of a radioactive tracer isotope is the source of the positrons used in positron emission tomography (PET scan).

### 9.1.3 Gamma Rays

A gamma ray, also known as gamma radiation (symbol $\gamma$), is a penetrating form of electromagnetic radiation arising from the radioactive decay of atomic nuclei. It consists of the shortest wavelength electromagnetic waves, typically shorter than those of X-rays. With frequencies above 30 exahertz ($3\times10^{19}$Hz) and wavelengths less than 10picometers ($1\times10^{-11}$m), gamma ray photons have the highest photon energy of any form of electromagnetic radiation. Paul Villard, a French chemist and physicist, discovered gamma radiation in 1900 while studying radiation emitted by radium. In 1903, Ernest Rutherford named this radiation *gamma rays* based on their relatively strong penetration of matter; in 1900, he had already named two less penetrating types of decay radiation (discovered by Henri Becquerel) alpha rays and beta rays in ascending order of penetrating power.

Gamma rays from radioactive decay are in the energy range from a few kiloelectronvolts (keV) to approximately 8 megaelectronvolts (MeV), corresponding to the typical energy levels in nuclei with reasonably long lifetimes. The energy spectrum of gamma rays can be used to identify the decaying radionuclides using gamma spectroscopy. Very-high-energy gamma rays in the 100–1000 teraelectronvolt (TeV) range have been observed from astronomical sources such as the Cygnus X-3 microquasar.

Natural sources of gamma rays originating on Earth are mostly a result of radioactive decay and secondary radiation from atmospheric interactions with cosmic ray particles. However, there are other rare natural sources, such as terrestrial gamma-ray flashes, which produce gamma rays from electron action upon the nucleus. Notable artificial sources of gamma rays include fission, such as that which occurs in nuclear reactors, and high energy physics experiments, such as neutral pion decay and nuclear fusion.

The energy ranges of gamma rays and X-rays overlap in the electromagnetic spectrum, so the terminology for these electromagnetic waves varies between scientific disciplines. In some fields of physics, they are distinguished by their origin: gamma rays are created by nuclear decay while X-rays originate outside the nucleus. In astrophysics, gamma rays are conventionally defined as having photon energies above 100 keV and are the subject of gamma-ray astronomy, while radiation below 100 keV is classified as X-rays and is the subject of X-ray astronomy.

Gamma rays are ionizing radiation and are thus hazardous to life. They can cause DNA mutations, cancer and tumors, and at high doses burns and radiation sickness. Due to their high penetration power, they can damage bone marrow and internal organs. Unlike alpha and beta rays, they easily pass through the body and thus pose a formidable radiation protection challenge, requiring shielding made from dense materials such as lead or concrete. On Earth, the magnetosphere protects life from most types of lethal cosmic radiation other than gamma rays.

## History of Discovery

The first gamma ray source to be discovered was the radioactive decay process called *gamma decay*. In this type of decay, an excited nucleus emits a gamma ray almost immediately upon formation. Paul Villard, a French chemist and physicist, discovered gamma radiation in 1900, while studying radiation emitted from radium. Villard knew that his described radiation was more powerful than previously described types of rays from radium, which included beta rays, first noted as "radioactivity" by Henri Becquerel in 1896, and alpha rays, discovered as a less penetrating form of radiation by Rutherford, in 1899.

However, Villard did not consider naming them as a different fundamental type. Later, in 1903, Villard's radiation was recognized as being of a type fundamentally different from previously named rays by Ernest Rutherford, who named Villard's rays "gamma rays" by analogy with the beta and alpha rays that Rutherford had differentiated in 1899. The "rays" emitted by radioactive elements were named in order of their power to penetrate various materials, using the first three letters of the Greek alphabet: alpha rays as the least penetrating, followed by beta rays, followed by gamma rays as the most penetrating. Rutherford also noted that gamma rays were not deflected (or at least, not *easily* deflected) by a magnetic field, another property making them unlike alpha and beta rays.

Gamma rays were first thought to be particles with mass, like alpha and beta rays. Rutherford initially believed that they might be extremely fast beta particles, but their failure to be deflected by a magnetic field indicated that they had no charge. In 1914, gamma rays were observed to be reflected from crystal surfaces, proving that they were electromagnetic radiation. Rutherford and his co-worker Edward Andrade measured the wavelengths of gamma rays from radium, and found they were similar to X-rays, but with shorter wavelengths and thus, higher frequency. This was eventually recognized as giving them more energy per photon, as soon as the latter term became generally accepted. A gamma decay was then understood to usually emit a gamma photon.

## Sources

Natural sources of gamma rays on Earth include gamma decay from naturally occurring radioisotopes such as potassium-40, and also as a secondary radiation from various atmospheric interactions with cosmic ray particles. Natural terrestrial sources that produce gamma rays include lightning strikes and terrestrial gamma-ray flashes, which produce high energy emissions from natural high-energy voltages. Gamma rays are produced by a number of astronomical processes in which very high-energy electrons are produced. Such electrons produce secondary gamma rays by the mechanisms of *bremsstrahlung*, inverse Compton scattering and synchrotron radiation. A large fraction of such astronomical gamma rays are screened by Earth's atmosphere. Notable artificial sources of gamma rays include fission, such as occurs in nuclear reactors, as well as high energy physics experiments, such as neutral pion decay and nuclear fusion.

A sample of gamma ray-emitting material that is used for irradiating or imaging is known as a gamma source. It is also called a radioactive source, isotope source, or radiation source, though these more general terms also apply to alpha and beta-emitting devices. Gamma sources are usually sealed to prevent radioactive contamination, and transported in heavy shielding.

## Radioactive Decay (Gamma Decay)

Gamma rays are produced during gamma decay, which normally occurs after other forms of decay occur, such as alpha or beta decay. A radioactive nucleus can decay by the emission of an $\alpha$ or $\beta$ particle. The daughter nucleus that results is usually left in an excited state. It can then decay to a lower energy state by emitting a gamma ray photon, in a process called gamma decay.

The emission of a gamma ray from an excited nucleus typically requires only $10^{-12}$ seconds. Gamma decay may also follow nuclear reactions such as neutron capture, nuclear fission, or nuclear fusion. Gamma decay is also a mode of relaxation of many excited states of atomic nuclei following other types of radioactive decay, such as beta decay, so long as these states possess the necessary component of nuclear spin.

When high-energy gamma rays, electrons, or protons bombard materials, the excited atoms emit characteristic "secondary" gamma rays, which are products of the creation of excited nuclear states in the bombarded atoms. Such transitions, a form of nuclear gamma fluorescence, form a topic in nuclear physics called gamma spectroscopy. Formation of fluorescent gamma rays are a rapid subtype of radioactive gamma decay.

In certain cases, the excited nuclear state that follows the emission of a beta particle or other type of excitation, may be more stable than average, and is termed a metastable excited state, if its decay takes (at least) 100 to 1000 times longer than the average $10^{-12}$ seconds. Such relatively long-lived excited nuclei are termed nuclear isomers, and their decays are termed isomeric transitions. Such nuclei have half-lifes that are more easily measurable, and rare nuclear isomers are able to stay in their excited state for minutes, hours, days, or occasionally far longer, before emitting a gamma ray.

The process of isomeric transition is therefore similar to any gamma emission, but differs in that it involves the intermediate metastable excited state(s) of the nuclei. Metastable states are often characterized by high nuclear spin, requiring a change in spin of several units or more with gamma decay, instead of a single unit transition that occurs in only $10^{-12}$ seconds. The rate of gamma decay is also slowed when the energy of excitation of the nucleus is small.

An emitted gamma ray from any type of excited state may transfer its energy directly to any electrons, but most probably to one of the K shell electrons of the atom, causing it to be ejected from that atom, in a process generally termed the photoelectric effect (external gamma rays and ultraviolet rays may also cause this effect).

The photoelectric effect should not be confused with the internal conversion process, in which a gamma ray photon is not produced as an intermediate particle (rather, a "virtual gamma ray" may be thought to mediate the process).

### Decay Schemes

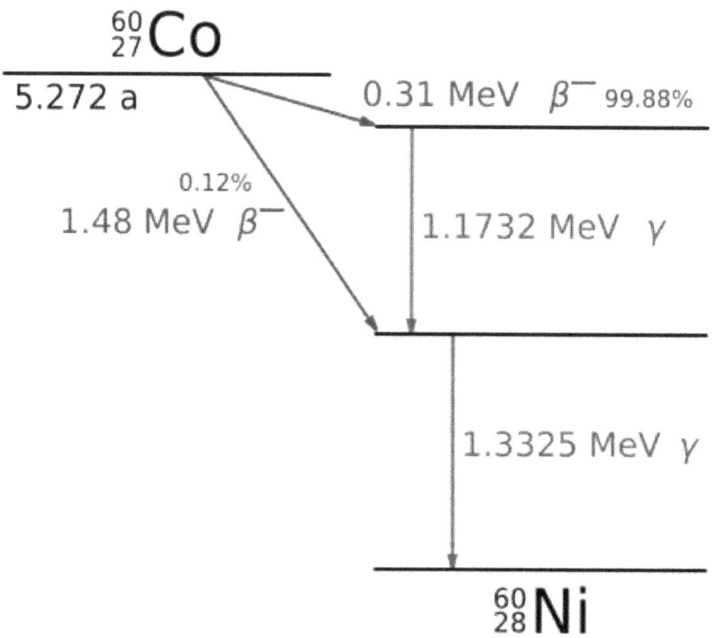


**Fig 8: Radioactive Decay Scheme of $^{60}Co$**

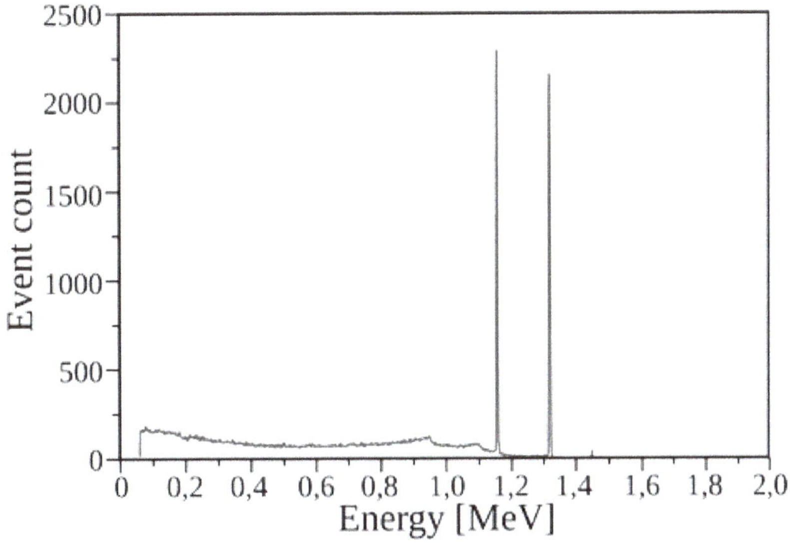


**Fig 9: Gamma Emission Spectrum of Cobalt-60**

One example of gamma ray production due to radionuclide decay is the decay scheme for cobalt-60, as illustrated in the accompanying diagram. First, $^{60}$Co decays to excited $^{60}$Niby beta decay emission of an electron of 0.31 MeV. Then the excited $^{60}$Nidecays to the ground state by emitting gamma rays in succession of 1.17 MeV followed by 1.33 MeV. This path is followed 99.88% of the time:

$$^{60}_{27}Co \rightarrow \; ^{60}_{28}Ni^{*} + \; e^{-} + \; \nu_{e} + \; \gamma + \; 1.17\,MeV$$

$$^{60}_{28}Ni^{*} \rightarrow \; ^{60}_{28}Ni \; + \; \gamma + \; 1.33\,MeV$$

Another example is the alpha decay of $^{241}$ Am to form $^{237}$ Np; which is followed by gamma emission. In some cases, the gamma emission spectrum of the daughter nucleus is quite simple, (e.g. $^{60}$Co/$^{60}$ Ni ) while in other cases, such as with ($^{241}$Am /$^{237}$Np and $^{192}$Ir/$^{192}$Pt ), the gamma emission spectrum is complex, revealing that a series of nuclear energy levels exist.

## Particle Physics

Gamma rays are produced in many processes of particle physics. Typically, gamma rays are the products of neutral systems which decay through electromagnetic interactions (rather than a weak or strong interaction). For example, in an electron–positron annihilation, the usual products are two gamma ray photons. If the annihilating electron and positron are at rest, each of the resulting gamma rays has an energy of ~ 511 keV and frequency of ~ $1.24 \times 10^{20}$ Hz. Similarly, a neutral pion most often decays into two photons. Many other hadrons and massive bosons also decay electromagnetically. High energy physics experiments, such as the Large Hadron Collider, accordingly employ substantial radiation shielding. Because subatomic particles mostly have far shorter wavelengths than atomic nuclei, particle physics gamma rays are generally several orders of magnitude more energetic than nuclear decay gamma rays. Since gamma rays are at the top of the electromagnetic spectrum in terms of energy, all extremely high-energy photons are gamma rays; for example, a photon having the Planck energy would be a gamma ray.

### Other Sources

A few gamma rays in astronomy are known to arise from gamma decay, but most do not. Photons from astrophysical sources that carry energy in the gamma radiation range are often explicitly called gamma-radiation. In addition to nuclear emissions, they are often produced by sub-atomic particle and particle-photon interactions. Those include electron-positron annihilation, neutral pion decay, bremsstrahlung, inverse Compton scattering, and synchrotron radiation.

- **Laboratory Sources:** In October 2017, scientists from various European universities proposed a means for sources of GeV photons using lasers as exciters through a controlled interplay between the cascade and anomalous radiative trapping.
- **Terrestrial Thunderstorms:** Thunderstorms can produce a brief pulse of gamma radiation called a terrestrial gamma-ray flash. These gamma rays are thought to be produced by high intensity static electric fields accelerating electrons, which then produce gamma rays by bremsstrahlung as they collide with and are slowed by atoms in the atmosphere. Gamma rays up to 100 MeV can be emitted by terrestrial thunderstorms, and were discovered by space-borne observatories. This raises the possibility of health risks to passengers and crew on aircraft flying in or near thunderclouds.
- **Solar Flares:** The most effusive solar flares emit across the entire EM spectrum, including $\gamma$-rays. The first confident observation occurred in 1972.
- Cosmic Rays: Extraterrestrial, high energy gamma rays include the gamma ray background produced when cosmic rays (either high speed electrons or protons) collide with ordinary matter, producing pair-production gamma rays at 511 keV. Alternatively, bremsstrahlung are produced at energies of tens of MeV or more when cosmic ray electrons interact with nuclei of sufficiently high atomic number.
- **Pulsars and Magnetars:** The gamma ray sky is dominated by the more common and longer-term production of gamma rays that emanate

from pulsars within the Milky Way. Sources from the rest of the sky are mostly quasars. Pulsars are thought to be neutron stars with magnetic fields that produce focused beams of radiation, and are far less energetic, more common, and much nearer sources than are quasars or the rarer gamma-ray burst sources of gamma rays. Pulsars have relatively long-lived magnetic fields that produce focused beams of relativistic speed charged particles, which emit gamma rays (bremsstrahlung) when those strike gas or dust in their nearby medium, and are decelerated. This is a similar mechanism to the production of high-energy photons in megavoltage radiation therapy machines. Inverse Compton scattering, in which charged particles (usually electrons) impart energy to low-energy photons boosting them to higher energy photons. Such impacts of photons on relativistic charged particle beams is another possible mechanism of gamma ray production. Neutron stars with a very high magnetic field (magnetars), thought to produce astronomical soft gamma repeaters, are another relatively long-lived star-powered source of gamma radiation.

**Fig 10: A Hypernova. Artist's Illustration Showing the Life of a Massive Star as Nuclear Fusion Converts Lighter Elements into Heavier Ones**

- **Quasars and Active Galaxies:** More powerful gamma rays from very distant quasars and closer active galaxies are thought to have a gamma ray production source similar to a particle accelerator. High energy electrons produced by the quasar, and subjected to inverse Compton scattering, synchrotron radiation, or bremsstrahlung, are the likely source of the gamma rays from those objects. It is thought that a supermassive black hole at the center of such galaxies provides the power source that intermittently destroys stars and focuses the resulting charged particles into beams that emerge from their rotational poles. When those beams interact with gas, dust, and lower energy photons they produce X-rays and gamma rays. These sources are known to fluctuate with durations of a few weeks, suggesting their relatively small size (less than a few light-weeks across). Such sources of gamma and X-rays are the most commonly visible high intensity sources outside the Milky Way galaxy. They shine not in bursts, but relatively continuously when viewed with gamma ray telescopes. The power of a typical quasar is about $10^{40}$ watts, a small fraction of which is gamma radiation. Much of the rest is emitted as electromagnetic waves of all frequencies, including radio waves.

When fusion no longer generates enough pressure to counteract gravity, the star rapidly collapses to form a black hole. Theoretically, energy may be released during the collapse along the axis of rotation to form a long duration gamma-ray burst.

- **Gamma-Ray Bursts:** The most intense sources of gamma rays, are also the most intense sources of any type of electromagnetic radiation presently known. They are the "long duration burst" sources of gamma rays in astronomy ("long" in this context, meaning a few tens of seconds). By contrast, "short" gamma-ray bursts of two seconds or less, which are not associated with supernovae, are thought to produce gamma rays during the collision of pairs of neutron stars, or a neutron star and a black hole. The so-called *long-duration* gamma-ray bursts produce a total energy output of about $10^{44}$ joules (as much energy as the Sun will produce in its entire life-time) but in a period of only 20 to 40 seconds. Gamma rays are approximately 50% of the total energy output. The leading hypotheses for the mechanism of production of these highest-known intensity beams of radiation, are inverse Compton scattering and synchrotron radiation from high-energy charged particles. These processes occur as relativistic charged particles leave the region of the event horizon of a newly formed black hole created during supernova explosion. The beam of particles moving at relativistic speeds are focused for a few tens of seconds by the magnetic field of the exploding hypernova. The fusion explosion of the hypernova drives the energetics of the process. If the narrowly directed beam happens to be pointed toward the Earth, it shines at gamma ray frequencies with such intensity, that it can be detected even at distances of up to 10 billion light years, which is close to the edge of the visible universe.

## Properties

### Penetration of Matter

Due to their penetrating nature, gamma rays require large amounts of shielding mass to reduce them to levels which are not harmful to living cells, in contrast to alpha particles, which can be stopped by paper or skin, and beta particles, which can be shielded by thin aluminium. Gamma rays are best absorbed by materials with high atomic numbers (Z) and high density, which contribute to the total stopping power. Because of this, a lead (high Z) shield is 20–30% better as a gamma shield than an equal mass of another low-Z shielding material, such as aluminium, concrete, water, or soil; lead's major advantage is not in lower weight, but rather its compactness due to its higher density. Protective clothing, goggles and respirators can protect from internal contact with or ingestion of alpha or beta emitting particles, but provide no protection from gamma radiation from external sources.

The higher the energy of the gamma rays, the thicker the shielding made from the same shielding material is required. Materials for shielding gamma rays are typically measured by the thickness required to reduce the intensity of the gamma rays by one half (the half-value layer or HVL). For example, gamma rays that require 1 cm (0.4 inch) of lead to reduce their intensity by 50% will also have their intensity reduced in half by 4.1 cm of granite rock, 6 cm (2.5 inches) of concrete, or 9 cm (3.5 inches) of packed soil. However, the mass of this much concrete or soil is only 20–30% greater than that of lead with the same absorption capability.

Depleted uranium is sometimes used for shielding in portable gamma ray sources, due to the smaller half-value layer when compared to lead (around 0.6 times the thickness for common gamma ray sources, i.e., Iridium-192 and Cobalt-60) and cheaper cost compared to tungsten.

In a nuclear power plant, shielding can be provided by steel and concrete in the pressure and particle containment vessel, while water provides a radiation shielding of fuel rods during storage or transport into the reactor core. The loss of water or removal of a "hot" fuel assembly into the air would result in much higher radiation levels than when kept under water.

High-energy (from 80 GeV to ~10 TeV) gamma rays arriving from far-distant quasars are used to estimate the extragalactic background light in the universe: The highest-energy rays interact more readily with the background light photons and thus the density of the background light may be estimated by analyzing the incoming gamma ray spectra.

Gamma spectroscopy is the study of the energetic transitions in atomic nuclei, which are generally associated with the absorption or emission of gamma rays. As in optical spectroscopy the absorption of gamma rays by a nucleus is especially likely (i.e., peaks in a "resonance") when the energy of the gamma ray is the same as that of an energy transition in the nucleus. In the case of gamma rays, such a resonance is seen in the technique of Mössbauer spectroscopy. In the Mössbauer effect the narrow resonance absorption for nuclear gamma absorption can be successfully attained by physically immobilizing atomic nuclei in a crystal. The immobilization of nuclei at both ends of a gamma resonance interaction is required so that no gamma energy is lost to the kinetic energy of recoiling nuclei at either the emitting or absorbing end of a gamma transition. Such loss of energy causes gamma ray resonance absorption to fail. However, when emitted gamma rays carry essentially all of the energy of the atomic nuclear de-excitation that produces them, this energy is also sufficient to excite the same energy state in a second immobilized nucleus of the same type.

## Applications

Gamma rays provide information about some of the most energetic phenomena in the universe; however, they are largely absorbed by the Earth's atmosphere. Instruments aboard high-altitude balloons and satellites missions, such as the Fermi Gamma-ray Space Telescope, provide our only view of the universe in gamma rays.

Gamma-induced molecular changes can also be used to alter the properties of semi-precious stones, and is often used to change white topaz into blue topaz.

Non-contact industrial sensors commonly use sources of gamma radiation in refining, mining, chemicals, food, soaps and detergents, and pulp and paper industries, for the measurement of levels, density, and thicknesses. Gamma-ray sensors are also used for measuring the fluid levels in water and oil industries. Typically, these use Co-60 or Cs-137 isotopes as the radiation source.

In the US, gamma ray detectors are beginning to be used as part of the Container Security Initiative (CSI). These machines are advertised to be able to scan 30 containers per hour.

Gamma radiation is often used to kill living organisms, in a process called irradiation. Applications of this include the sterilization of medical equipment (as an alternative to autoclaves or chemical means), the removal of decay-causing bacteria from many foods and the prevention of the sprouting of fruit and vegetables to maintain freshness and flavor.

Despite their cancer-causing properties, gamma rays are also used to treat some types of cancer, since the rays also kill cancer cells. In the procedure called gamma-knife surgery, multiple concentrated beams of gamma rays are directed to the growth in order to kill the cancerous cells. The beams are aimed from different angles to concentrate the radiation on the growth while minimizing damage to surrounding tissues.

Gamma rays are also used for diagnostic purposes in nuclear medicine in imaging techniques. A number of different gamma-emitting radioisotopes are used. For example, in a PET scan a radiolabeled sugar called fluorodeoxyglucose emits positrons that are annihilated by electrons, producing pairs of gamma rays that highlight cancer as the cancer often has a higher metabolic rate than the surrounding tissues. The most common gamma emitter used in medical applications is the nuclear isomer technetium-99m which emits gamma rays in the same energy range as diagnostic X-rays. When this radionuclide tracer is administered to a patient, a gamma camera can be used to form an image of the radioisotope's distribution by detecting the gamma radiation emitted. Depending on which molecule has been labeled with the tracer, such techniques can be employed to diagnose a wide range of conditions (for example, the spread of cancer to the bones via bone scan).

## Health Effects

Gamma rays cause damage at a cellular level and are penetrating, causing diffuse damage throughout the body. However, they are less ionising than alpha or beta particles, which are less penetrating.

Low levels of gamma rays cause a stochastic health risk, which for radiation dose assessment is defined as the *probability* of cancer induction and genetic damage. The International Commission on Radiological Protection says "In the low dose range, below about 100 mSv, it is scientifically plausible to assume that the incidence of cancer or heritable effects will rise in direct proportion to an increase in the equivalent dose in the relevant organs and tissues" High doses produce deterministic effects, which is the *severity* of acute tissue damage that is certain to happen. These effects are compared to the physical quantity absorbed dose measured by the unit gray (Gy).

When gamma radiation breaks DNA molecules, a cell may be able to repair the damaged genetic material, within limits. However, a study of Rothkamm and Lobrich has shown that this repair process works well after high-dose exposure but is much slower in the case of a low-dose exposure.

Studies have shown low-dose gamma radiation may be enough to cause cancer. In a study of mice, they were given human-relevant low-dose gamma radiation, with genotoxic effects 45 days after continuous low-dose gamma radiation, with significant increases of chromosomal damage, DNA lesions and phenotypic mutations in blood cells of irradiated animals, covering the three types of genotoxic activity. Another study studied the effects of acute ionizing gamma radiation in rats, up to 10 Gy, and who ended up showing acute oxidative protein damage, DNA damage, cardiac troponin T carbonylation, and long-term cardiomyopathy.

The natural outdoor exposure in the United Kingdom ranges from 0.1 to 0.5 μSv/h with significant increase around known nuclear and contaminated sites. Natural exposure to gamma rays is about 1 to 2 mSv per year, and the average total amount of radiation received in one year per inhabitant in the USA is 3.6 mSv. There is a small increase in the dose, due to naturally occurring gamma radiation, around small particles of high atomic number materials in the human body caused by the photoelectric effect.

By comparison, the radiation dose from chest radiography (about 0.06 mSv) is a fraction of the annual naturally occurring background radiation dose. A chest CT delivers 5 to 8 mSv. A whole-body PET/CT scan can deliver 14 to 32 mSv depending on the protocol. The dose from fluoroscopy of the stomach is much higher, approximately 50 mSv (14 times the annual background).

An acute full-body equivalent single exposure dose of 1 Sv (1000 mSv), or 1 Gy, will cause mild symptoms of acute radiation sickness, such as nausea and vomiting; and a dose of 2.0–3.5 Sv (2.0–3.5 Gy) causes more severe symptoms (i.e. nausea, diarrhea, hair loss, hemorrhaging, and inability to fight infections), and will cause death in a sizable number of cases—about 10% to 35% without medical treatment. A dose of 3–5 Sv (3–5 Gy) is considered approximately the $LD_{50}$ (or the lethal dose for 50% of exposed population) for an acute exposure to radiation even with standard medical treatment. A dose higher than 5 Sv (5 Gy) brings an increasing chance of death above 50%. Above 7.5–10 Sv (7.5–10 Gy) to the entire body, even extraordinary treatment, such as bone-marrow transplants, will not prevent the death of the individual exposed. (Doses much larger than this may, however, be delivered to selected parts of the body in the course of radiation therapy.)

For low-dose exposure, for example among nuclear workers, who receive an average yearly radiation dose of 19 mSv, the risk of dying from cancer (excluding leukemia) increases by 2 percent. For a dose of 100 mSv, the risk increase is 10 percent. By comparison, risk of dying from cancer was increased by 32 percent for the survivors of the atomic bombing of Hiroshima and Nagasaki.

## Distinction from X-Rays

The conventional distinction between X-rays and gamma rays has changed over time. Originally, the electromagnetic radiation emitted by X-ray tubes almost invariably had a longer wavelength than the radiation (gamma rays) emitted by radioactive nuclei. Older literature distinguished between X- and gamma radiation on the basis of wavelength, with radiation shorter than some arbitrary wavelength, such as $10^{-11}$ m, defined as gamma rays. Since the energy of photons is proportional to their frequency and inversely proportional to wavelength, this past distinction between X-rays and gamma rays can also be thought of in terms of its energy, with gamma rays considered to be higher energy electromagnetic radiation than are X-rays.

However, since current artificial sources are now able to duplicate any electromagnetic radiation that originates in the nucleus, as well as far higher energies, the wavelengths characteristic of radioactive gamma ray sources vs. other types now completely overlap. Thus, gamma rays are now usually distinguished by their origin: X-rays are emitted by definition by electrons outside the nucleus, while gamma rays are emitted by the nucleus. Exceptions to this convention occur in astronomy, where gamma decay is seen in the afterglow of certain supernovas, but radiation from high energy processes known to involve other radiation sources than radioactive decay is still classed as gamma radiation.

For example, modern high-energy X-rays produced by linear accelerators for megavoltage treatment in cancer often have higher energy (4 to 25 MeV) than do most classical gamma rays produced by nuclear gamma decay. One of the most common gamma ray emitting isotopes used in diagnostic nuclear medicine, technetium-99m, produces gamma radiation of the same energy (140 keV) as that produced by diagnostic X-ray machines, but of significantly lower energy than therapeutic photons from linear particle accelerators. In the medical community today, the convention that radiation produced by nuclear decay is the only type referred to as "gamma" radiation is still respected.

Due to this broad overlap in energy ranges, in physics the two types of electromagnetic radiation are now often defined by their origin: X rays are emitted by electrons (either in orbitals outside of the nucleus, or while being accelerated to produce bremsstrahlung-type radiation), while gamma rays are emitted by the nucleus or by means of other particle decays or annihilation events. There is no lower limit to the energy of photons produced by nuclear reactions, and thus ultraviolet or lower energy photons produced by these processes would also be defined as "gamma rays" (indeed, this happens for the isomeric transition of the extremely low-energy isomer $^{229m}$Th).

The only naming-convention that is still universally respected is the rule that electromagnetic radiation that is known to be of atomic nuclear origin is *always* referred to as "gamma rays", and never as X-rays. However, in physics and astronomy, the converse convention (that all gamma rays are considered to be of nuclear origin) is frequently violated.

In astronomy, higher energy gamma and X-rays are defined by energy, since the processes that produce them may be uncertain and photon energy, not origin, determines the required astronomical detectors needed. High-energy photons occur in nature that are known to be produced by processes other than nuclear decay but are still referred to as gamma radiation. An example is "gamma rays" from lightning discharges at 10 to 20 MeV, and known to be produced by the bremsstrahlung mechanism.

Another example is gamma-ray bursts, now known to be produced from processes too powerful to involve simple collections of atoms undergoing radioactive decay. This is part and parcel of the general realization that many gamma rays produced in astronomical processes result not from radioactive decay or particle annihilation, but rather in non-radioactive processes similar to X-rays. Although the gamma rays of astronomy often come from non-radioactive events, a few gamma rays in astronomy are specifically known to originate from gamma decay of nuclei (as demonstrated by their spectra and emission half life).

A classic example is that of supernova SN 1987A, which emits an "afterglow" of gamma-ray photons from the decay of newly made radioactive nickel-56 and cobalt-56. Most gamma rays in astronomy, however, arise by other mechanisms.

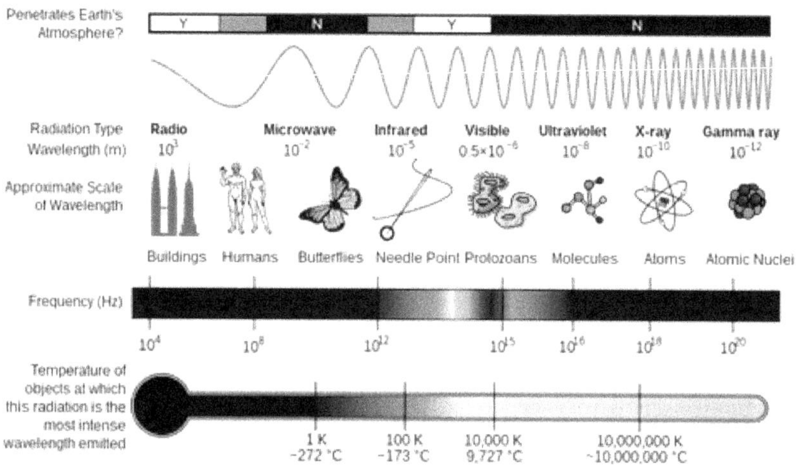


**Fig 11: In Practice, Gamma Ray Energies Overlap with the Range of X-Rays, Especially in the Higher-Frequency Region Referred to as "Hard" X-Rays.**

## 9.2 COMMONLY USED RADIOISOTOPES

Several radioisotopes are widely used due to their stable decay rates and compatibility with biological molecules.

- **Carbon-14 (C-14):** Carbon-14 is a beta emitter with a long half-life, making it ideal for studying metabolic pathways, such as carbohydrate or protein metabolism, in long-term studies.
- **Tritium (H-3):** Tritium is a low-energy beta emitter used in biochemical research to label compounds and study DNA, RNA, and protein synthesis.
- **Iodine-125 (I-125):** Iodine-125 is a gamma emitter often used in immunoassays and radioimmunoassays, particularly for tracking thyroid activity and protein binding studies.
- **Technetium-99m (Tc-99m):** With a short half-life, Technetium-99m is a popular choice in medical diagnostics for imaging organs such as the heart, liver, and lungs.
- **Phosphorus-32 (P-32):** Phosphorus-32 is a beta emitter widely used in

molecular biology to study nucleic acids and phosphorylated molecules in cellular processes.

These radioisotopes serve specific functions based on their decay properties, allowing researchers to choose appropriate isotopes for tracking and analyzing distinct biochemical pathways.

## 9.3 APPLICATIONS OF RADIOISOTOPE TECHNIQUES

Radioisotope techniques have revolutionized research in various fields, particularly in the study of biological, chemical, and medical processes.

### 9.3.1 Medical Diagnostics and Imaging

Radioisotopes play a crucial role in diagnostic imaging by allowing doctors to visualize internal organs and detect abnormalities. Techniques such as Positron Emission Tomography (PET) and Single Photon Emission Computed Tomography (SPECT) rely on radioisotopes like fluorine-18 and technetium-99m, respectively, to create detailed images of the body's organs and tissues. These techniques are instrumental in diagnosing and monitoring diseases such as cancer, heart disease, and neurological disorders. For instance, PET scans can detect tumors by tracking the accumulation of radioactive glucose, as cancerous cells absorb glucose at a higher rate than normal cells.

### 9.3.2 Cancer Treatment

Radioisotopes are also used in cancer treatment, a method known as radiotherapy. Radioactive isotopes such as iodine-131 and radium-223 are used to target and destroy cancerous cells. Iodine-131 is specifically effective in treating thyroid cancer, as the thyroid naturally absorbs iodine, allowing targeted treatment. In brachytherapy, radioactive seeds containing isotopes like iridium-192 are implanted near the tumor to deliver localized radiation, minimizing damage to surrounding tissues. This targeted approach improves the efficacy of treatment and reduces side effects.

### 9.3.3 Biological Research

In biological research, radioisotope techniques have made it possible to study complex metabolic pathways, enzyme kinetics, and genetic processes. For example, carbon-14 and tritium-labeled compounds are commonly used in studying cellular processes such as DNA replication, RNA transcription, and protein synthesis.

By incorporating radioisotopes into specific biomolecules, researchers can track their movement, measure rates of synthesis or degradation, and identify intermediates within biochemical pathways. This has led to advancements in understanding metabolic disorders, hormonal regulation, and molecular biology.

### 9.3.4 Environmental Studies

Radioisotope techniques are also applied in environmental studies to trace pollutant movement and measure ecosystem changes. Radioisotopes such as strontium-90 and cesium-137 are used to track soil erosion, water contamination, and the spread of radioactive materials in the environment. These techniques help monitor pollution sources, study biogeochemical cycles, and assess environmental impact, providing valuable insights for conservation efforts and environmental protection.

## 9.4 TECHNIQUES AND METHODOLOGIES IN RADIOISOTOPE STUDIES

The practical application of radioisotope techniques involves a range of specialized methodologies that allow precise measurements and analyses.

### 9.4.1 Autoradiography

Autoradiography is a technique where radioactively labeled compounds are detected on a photographic film. After incorporating a radioisotope into a biological sample, the sample is placed on a photographic film, where radiation exposure reveals the presence and distribution of the radioactive compound. Autoradiography is widely used in cell biology to study gene expression, protein localization, and metabolic pathways.

### 9.4.2 Scintillation Counting

Scintillation counting is a method for measuring radioactivity in a sample by detecting light pulses produced by radioactive decay. The sample is mixed with a scintillation solution, which emits light when struck by radiation. A scintillation counter detects these light pulses and measures the intensity of radioactivity. This technique is highly sensitive and commonly used in studies of enzyme kinetics, protein interactions, and nucleic acid hybridization.

### 9.4.3 Radioimmunoassay

Radioimmunoassay (RIA) is a technique used to measure antigen concentrations, such as hormones or proteins, in a sample. It involves binding a radioactively labeled antigen to its specific antibody and measuring the bound versus unbound antigen. The amount of radioactivity indicates the antigen concentration in the sample. RIA is widely used in clinical diagnostics to measure hormones, drugs, and other biomolecules.

### 9.4.4 Tracer Studies

Tracer studies involve the incorporation of radioisotopes into molecules, allowing researchers to track their movement and measure reaction rates in biological or chemical processes. Tracer studies are instrumental in metabolism research, enabling scientists to identify intermediates in metabolic pathways and understand energy flow within cells.

## 9.5 SAFETY AND ETHICAL CONSIDERATIONS

The use of radioisotopes poses several safety and ethical challenges due to the risks associated with radiation exposure. Proper handling, disposal, and storage of radioactive materials are crucial to minimize health risks to researchers and the environment. Strict guidelines and regulations govern the use of radioisotopes in laboratories, ensuring that exposure to radiation is kept within safe limits.

- **Personal Protective Equipment (PPE):** Researchers handling radioisotopes are required to use PPE, such as lab coats, gloves, and radiation shields, to minimize exposure. Work is often conducted in controlled environments, such as fume hoods or shielded rooms, to protect personnel and limit radiation spread.
- **Ethical Considerations:** Radioisotope techniques are also subject to ethical scrutiny, particularly in clinical applications where patient exposure to radiation is involved. Ethical guidelines ensure that the benefits of using radioisotopes outweigh the risks, with a focus on minimizing radiation dose and providing informed consent for diagnostic or therapeutic procedures.

## 9.6 ADVANTAGES AND LIMITATIONS OF RADIOISOTOPE TECHNIQUES

Radioisotope techniques offer several advantages but also have limitations.

### 9.6.1 Advantages

- **High Sensitivity:** Radioisotope techniques are highly sensitive, allowing the detection and quantification of minute amounts of substances.
- **Specificity:** Radioisotopes can be incorporated into specific molecules, enabling targeted studies of metabolic pathways and molecular interactions.
- **Quantitative Analysis:** These techniques allow precise measurement of reaction rates, molecular concentrations, and cellular activities.

### 9.6.2 Limitations

- **Radiation Hazard:** Radioisotopes pose health risks to researchers and the environment if not handled correctly.
- **Cost and Availability:** The procurement and disposal of radioisotopes require significant resources, limiting accessibility.

- **Short Half-Life:** Some radioisotopes have short half-lives, requiring quick processing and analysis, which may be challenging in some studies.

## 9.7 FUTURE OF RADIOISOTOPE TECHNIQUES

The future of radioisotope techniques is closely tied to advances in technology and safety protocols. Efforts to improve radiation safety and reduce the environmental impact of radioactive waste are essential for the continued use of radioisotopes in research. Innovations in imaging, such as hybrid techniques that combine radioisotopes with fluorescent tags, promise to enhance the specificity and resolution of studies.

In medical diagnostics, advancements in radioisotope tracers may lead to more personalized and effective treatments, particularly in oncology. By targeting specific biomarkers with radioisotopes, doctors may be able to deliver treatments more precisely, reducing side effects and improving patient outcomes.

## 9.8 CONCLUSION

Radioisotope techniques have transformed the way researchers and clinicians study biological, chemical, and environmental processes. The ability to trace molecules at the atomic level has provided unprecedented insights into cellular function, disease mechanisms, and ecological changes. Despite challenges related to safety and cost, radioisotope techniques remain indispensable in modern science. With ongoing improvements in safety and technology, radioisotope applications are likely to expand, further advancing fields such as medicine, environmental science, and biochemistry.

# CHAPTER 10

# PRINCIPLES OF CLINICAL BIOCHEMISTRY

Clinical biochemistry is a branch of medical science that involves the analysis of bodily fluids for diagnostic and therapeutic purposes. It plays a crucial role in healthcare by providing insights into physiological and biochemical abnormalities that signal disease states.

The principles of clinical biochemistry guide how laboratories collect, measure, and interpret biochemical data from blood, urine, and other fluids. These principles ensure accuracy and reliability in testing, helping healthcare providers make informed decisions on patient care.

## 10.1 IMPORTANCE OF CLINICAL BIOCHEMISTRY IN MEDICINE

Clinical biochemistry is integral to diagnosing and monitoring diseases. Through biochemical testing, clinicians can detect metabolic imbalances, identify organ dysfunction, and monitor the effectiveness of treatments. For example, tests measuring blood glucose levels are essential for diagnosing and managing diabetes, while liver function tests provide information on liver health. By assessing enzyme activities, hormone levels, and electrolyte concentrations, clinical biochemistry aids in diagnosing a wide range of conditions, from metabolic disorders to cardiovascular diseases.

## 10.2 BASIC ANALYTICAL TECHNIQUES

Clinical biochemistry relies on various analytical techniques to measure the chemical components of bodily fluids. These methods include spectrophotometry, electrophoresis, chromatography, and immunoassays, each chosen for their ability to analyze specific biochemical markers with precision.

### 10.2.1 Spectrophotometry

Spectrophotometry is one of the most widely used techniques in clinical biochemistry. It measures the concentration of substances by analyzing the absorption or emission of light at specific wavelengths. For example, spectrophotometry is used to measure hemoglobin levels in blood, providing a quick and reliable assessment of a patient's oxygen-carrying capacity.

### 10.2.2 Electrophoresis

Electrophoresis separates molecules based on their charge and size by applying an electric field. It is commonly used in protein and DNA analysis. Serum protein electrophoresis, for instance, can help diagnose diseases such as multiple myeloma by identifying abnormal proteins in the blood.

### 10.2.3 Chromatography

Chromatography techniques, such as gas chromatography (GC) and high-performance liquid chromatography (HPLC), separate compounds based on their chemical properties. These methods are essential in toxicology and pharmacology, where precise measurement of drug metabolites or toxins in blood or urine is required.

### 10.2.4 Immunoassays

Immunoassays use antibodies to detect specific biomolecules, making them particularly useful for measuring hormone levels, such as thyroid hormones, or detecting antigens associated with infections. Enzyme-linked immunosorbent assay (ELISA) is a common type of immunoassay used in clinical settings for various tests, including HIV and hepatitis screenings.

## 10.3 STANDARDIZATION AND CALIBRATION

Standardization and calibration are fundamental principles in clinical biochemistry to ensure test accuracy and reliability. These processes involve using standards, controls, and calibration methods to verify that instruments and reagents provide consistent results.

### 10.3.1 Standards and Controls

Standards are solutions with known concentrations of analytes used to calibrate instruments and establish baseline values for tests. Controls, on the other hand, are samples with known properties used to verify the accuracy of test procedures. Using both standards and controls allows laboratories to detect potential errors in test results and maintain consistent measurements over time.

### 10.3.2 Calibration of Instruments

Instrument calibration is essential to accurate testing, as it ensures that the equipment measures concentrations within an acceptable range of accuracy. Regular calibration of spectrophotometers, analyzers, and other devices used in clinical biochemistry helps prevent measurement errors that could lead to misdiagnosis or incorrect treatment.

## 10.4 QUALITY CONTROL AND QUALITY ASSURANCE

Quality control (QC) and quality assurance (QA) are systematic processes used to maintain high standards in clinical biochemistry testing. QC involves daily checks and procedures that validate test accuracy, while QA encompasses the overall processes and policies in place to ensure consistent quality.

### 10.4.1 Internal Quality Control

Internal quality control involves the daily monitoring of tests to detect any deviations from expected results. This includes the use of control samples to identify discrepancies within individual tests, such as blood glucose or cholesterol measurements. By spotting errors early, laboratories can take corrective action to prevent inaccurate reporting.

### 10.4.2 External Quality Assessment

External Quality Assessment (EQA) involves comparing results between different laboratories to ensure consistency and reliability across facilities. EQA programs provide blind samples to participating labs, allowing an external agency to evaluate their performance. This helps to identify discrepancies and improve testing practices on a broader scale.

### 10.4.3 Quality Assurance Programs

Quality assurance programs in clinical biochemistry ensure that procedures, equipment, and personnel meet industry standards. These programs cover all aspects of the testing process, from sample collection to reporting, and include regular audits, staff training, and documentation. QA programs ensure compliance with regulatory bodies and accreditation standards, promoting patient safety and confidence in test results.

## 10.5 CLINICAL APPLICATIONS OF BIOCHEMICAL TESTING

Clinical biochemistry testing is used to diagnose, monitor, and manage a wide variety of diseases, including metabolic disorders, kidney disease, liver disease, and cardiovascular disease. Biochemical markers, such as glucose, lipids, enzymes, and electrolytes, provide valuable insights into a patient's health.

### 10.5.1 Diagnosis of Metabolic Disorders

Tests measuring blood glucose, lipid profiles, and enzyme levels are essential for diagnosing metabolic disorders like diabetes and hyperlipidemia. These tests help in identifying risk factors early, allowing for lifestyle changes or treatments to prevent disease progression.

### 10.5.2 Monitoring of Chronic Conditions

Clinical biochemistry is critical in managing chronic conditions. For example, routine blood tests to measure glycated hemoglobin (HbA1c) levels are important for diabetes management, as they provide a long-term overview of blood glucose control. Similarly, liver function tests are used to monitor liver health in patients with chronic hepatitis or cirrhosis.

### 10.5.3 Cardiovascular Risk Assessment

Biochemical markers such as cholesterol, triglycerides, and high-sensitivity C-reactive protein (hs-CRP) are used in assessing cardiovascular disease risk. Elevated levels of these markers indicate a higher risk of heart disease, prompting lifestyle interventions or medical treatments to manage the risk factors.

### 10.5.4 Cancer Detection and Monitoring

Tumor markers, such as prostate-specific antigen (PSA) and carcinoembryonic antigen (CEA), play a role in diagnosing and monitoring certain cancers. While not definitive, these markers provide helpful information in conjunction with imaging and biopsy results, allowing for timely intervention and treatment adjustments.

## 10.6 INTERPRETATION OF BIOCHEMICAL RESULTS

Accurate interpretation of biochemical results is essential for diagnosis and treatment planning. Factors like age, sex, diet, and underlying health conditions influence reference ranges for biochemical markers, making context crucial in result interpretation.

### 10.6.1 Reference Ranges

Reference ranges are established values for biochemical markers within which most healthy individuals fall. These ranges can vary by population, and clinical biochemists consider factors such as age, gender, and lifestyle when interpreting results. A value outside the reference range may indicate a potential health issue but must be interpreted within the clinical context.

### 10.6.2 Sensitivity and Specificity of Tests

The sensitivity and specificity of a test determine its effectiveness in diagnosing a condition. A highly sensitive test is effective in identifying those with a disease, while a highly specific test accurately excludes those without the disease. Balancing these characteristics is essential in clinical biochemistry to minimize false positives and false negatives.

## 10.7 ETHICAL AND SAFETY CONSIDERATIONS

Clinical biochemistry involves ethical and safety considerations related to patient confidentiality, sample handling, and result reporting. Maintaining ethical standards in clinical biochemistry ensures that patient information is handled responsibly and securely.

### 10.7.1 Patient Confidentiality

Patient confidentiality is paramount in clinical biochemistry. Laboratories follow strict guidelines to ensure that patient information is safeguarded, allowing only authorized personnel access to sensitive data.

### 10.7.2 Safety in Sample Handling

The handling of biological samples poses risks, including exposure to infectious agents. Laboratory personnel adhere to safety protocols, including the use of personal protective equipment (PPE) and proper disposal procedures, to protect themselves and others.

## 10.8 FUTURE DIRECTIONS IN CLINICAL BIOCHEMISTRY

Advances in technology continue to shape the future of clinical biochemistry. Innovations such as point-of-care testing, molecular diagnostics, and the use of artificial intelligence in data analysis offer potential for faster, more accurate, and personalized diagnostics.

### 10.8.1 Point-of-Care Testing

Point-of-care testing (POCT), also called near-patient testing or bedside testing, is defined as medical diagnostic testing at or near the point of care—that is, at the time and place of patient care. This contrasts with the historical pattern in which testing was wholly or mostly confined to the medical laboratory, which entailed sending off specimens away from the point of care and then waiting hours or days to learn the results, during which time care must continue without the desired information.

#### Technology Overview

Point-of-care tests are simple medical tests that can be performed at the bedside. In many cases, the simplicity was not achievable until technology developed not only to make a test possible at all but then also to mask its complexity.

For example, various kinds of urine test strips have been available for decades, but portable ultrasonography did not reach the stage of being advanced, affordable, and widespread until the 2000s and 2010s.

Today, portable ultrasonography is often viewed as a "simple" test, but there was nothing simple about it until the more complex technology was available. Similarly, pulse oximetry can test arterial oxygen saturation in a quick, simple, noninvasive, affordable way today, but in earlier eras this required an intra-arterial needle puncture and a laboratory test; and rapid diagnostic tests such as malaria antigen detection tests or COVID-19 rapid tests that rely on a state of the art in immunology that did not exist until recent decades.

Thus, over decades, testing continues to move toward the point of care more than it formerly had been. A recent survey in five countries (Australia, Belgium, the Netherlands, the UK and the US) indicates that general practitioners / family doctors would like to use more POCTs.

The driving notion behind POCT is to bring the test conveniently and immediately to the patient. This increases the likelihood that the patient, physician, and care team will receive the results quicker, which allows for better immediate clinical management decisions to be made. POCT includes: blood glucose testing, blood gas and electrolytes analysis, rapid coagulation testing, rapid cardiac markers diagnostics, drugs of abuse screening, urine strips testing, pregnancy testing, fecal occult blood analysis, food pathogens screening, hemoglobin diagnostics, infectious disease testing (such as COVID-19 rapid tests), cholesterol screening and emerging technologies in micronutrient deficiency screening and diagnosis of acute febrile illness.

Lab-on-a-chip technologies are one of the main drivers of point-of-care testing, especially in the field of infectious disease diagnosis. These technologies enable different bioassays such as microbiological culture, PCR, ELISA to be used at the point of care.

POCT is often accomplished through the use of transportable, portable, and handheld instruments (e.g., blood glucose meter, nerve conduction study device) and test kits (e.g., CRP, HBA1C, Homocystein, HIV salivary assay, etc.). Small bench analyzers or fixed equipment can also be used when a handheld device is not available—the goal is to collect the specimen and obtain the results in a very short period of time at or near the location of the patient so that the treatment plan can be adjusted as necessary before the patient leaves. Cheaper, faster, and smarter POCT devices have increased the use of POCT approaches by making it cost-effective for many diseases, such as diabetes, carpal tunnel syndrome (CTS) and acute coronary syndrome. Additionally, it is very desirable to measure various analytes simultaneously in the same specimen, allowing a rapid, low-cost, and reliable quantification. Therefore, multiplexed point-of-care testing (xPOCT) has become more important for medical diagnostics in the last decade.

Many point-of-care test systems are realized as easy-to-use membrane-based test strips, often enclosed by a plastic test cassette. This concept often is realized in test systems for detecting pathogens, the most common being COVID-19 rapid tests. Very recently such test systems for rheumatology diagnostics have been developed, too. These tests require only a single drop of whole blood, urine or saliva, and they can be performed and interpreted by any general physician within minutes. Recently, a portable medical diagnostic device called "BioPoC" has been reported which employs free-standing enzyme-modified responsive polymer membrane-based biosensors and a newly devised low-cost transduction principle for the detection of *H. pylori* and urea.

During the COVID-19 pandemic, rapid development of POCT occurred, aiming to improve the turnaround time and ease of use compared to the gold standard lab-based PCR test. These have included rapid antigen tests, alternate nucleic acid amplification methods, and novel sensors. A range of test have been developed including smartphone based platforms, and tests targeting blood, saliva, faecal matter, urine, and tears have been proposed. Saliva in particular may offer sufficiently high detection rates in tandem with a non-invasive and user friendly procedure, although reliability requires improvement.

Emerging technology at the point of care setting is being developed to allow for rapid assessment of micronutrient deficiency. The Cornell NutriPhone is a promising technology for determining nutritional status at the point of care. This technology allows assessment of iron, vitamin A, vitamin D, and vitamin B12 from a single drop of blood in around 15 minutes. Building on this same platform, there are proof-of-concept studies for fever and cancer.

## Benefits

The coupling of POCT devices and electronic medical records enable test results to be shared instantly with care providers. The use of mobile devices in the health care setting also enable the health care provider to quickly access patient test results sent from a POCT device. A reduction in morbidity and mortality has been associated with such rapid turn around times from a study using the i-STAT to analyze blood lactate levels after congenital heart surgery.

POCT has become established worldwide and finds vital roles in public health. Many researchers emphasize POCT as the normal standard of care in disaster situations.

Potential operational benefits include more rapid decision making and triage, reduced operating times, high-dependency, postoperative care time, emergency room time, number of outpatient clinic visits, number of hospital beds required, ensuring optimal use of professional time and reduced of antimicrobial medication.

At home or POCT tests, providing results within minutes of being administered, would allow for appropriate measures and rapid decisions about dental patients' care process. Characteristics and detection rate of SARS-CoV-2 in alternative sites and specimens related to dentistry has been extensively reviewed.

## Regulatory in the U.S.

The Clinical Laboratory Improvement Amendments (CLIA) regulate any laboratory testing and require laboratories to obtain certificates to do any testing on human specimens for health assessment or to diagnose, prevent, or treat disease. Three federal agencies partner together to cover the responsibilities put forward in the regulations: the Food and Drug Administration (FDA), Centers for Medicare & Medicaid Services (CMS), and the Centers for Disease Control and Prevention (CDC). In addition, accreditation standards developed by Joint Commission and College of American Pathologists have been developed to improve quality.

### Food and Drug Administration (FDA)

In vitro diagnostic (IVD) products use the same categorization as medical devices (Class I, II, and III) to assure safety and effectiveness. Regulatory controls and premarket approval process are determined by this classification, with Class I being the lowest risk (least regulated) and Class III being the highest risk (most regulated).

Under the CLIA, it is the role of the FDA to assess the complexity of the in vitro laboratory diagnostic tests. Tests are only scored after the FDA has cleared or approved a premarketing request, or upon request. Manufacturers can apply for CLIA waivers during this premarket approval/clearance process. Tests that are already cleared or approved for home use or are waived by 42 CRF 293.15(c), are classified as waived. Otherwise, the tests are either classified as moderate or high complexity based on seven categorization criteria listed in 42 CFR 493.17. If the test is classified as moderate, the manufacturer may request the test be waived through the CLIA Waiver by Application. The application must show that the test meets the criteria in 42 U.S.C. § 263a(d)(3), that the test is simple and will not cause harm to the patient if performed incorrectly.

These test classifications determine the certifications needed for laboratories to perform said tests. Waived tests require the least regulation, while moderate to high complexity tests require higher regulation and standards within the laboratory.

### Center for Medicaid Services (CMS)

Under CLIA, it is the role of CMS to issue laboratory certificates and monitor, inspect, and enforce laboratory regulatory compliance based on the tests being performed. In total, CMS covers 260,000 laboratories.

### Centers for Disease Control and Prevention (CDC)

The CDC focuses on the analysis, research, and technical assistance within the CLIA partnership. In particular, the CDC establishes technical standards and guidelines, conducting studies, monitoring practices, and developing resources. In addition, the CDC manages the Clinical Laboratory Improvement Advisory Committee (CLIAC). CLIAC is made up of experts in many specialties throughout clinical and anatomic pathology that provide guidance and advice on general issues within laboratory science. The CDC specifically acknowledges that point-of-care testing simply describes the location at which the testing is performed and not the complexity of the test itself. With technological innovation, more complex tests will bc able to be performed at the bedside that may not be CLIA-waived like some other at-home point of care tests that the FDA has waived such as urine dipsticks.

## 10.8.2 Molecular Diagnostics

Molecular diagnostics is a collection of techniques used to analyze biological markers in the genome and proteome, and how their cells express their genes as proteins, applying molecular biology to medical testing. In medicine the technique is used to diagnose and monitor disease, detect risk, and decide which therapies will work best for individual patients, and in agricultural biosecurity similarly to monitor crop- and livestock disease, estimate risk, and decide what quarantine measures must be taken.

By analysing the specifics of the patient and their disease, molecular diagnostics offers the prospect of personalised medicine. These tests are useful in a range of medical specialties, including infectious disease, oncology, human leucocyte antigen typing (which investigates and predicts immune function), coagulation, and pharmacogenomics—the genetic prediction of which drugs will work best. They overlap with clinical chemistry (medical tests on bodily fluids).

### History

The field of molecular biology grew in the late twentieth century, as did its clinical application. In 1980, Yuet Wai Kan *et al.* suggested a prenatal genetic test for Thalassemia that did not rely upon DNA sequencing—then in its infancy—but on restriction enzymes that cut DNA where they recognised specific short sequences, creating different lengths of DNA strand depending on which allele (genetic variation) the fetus possessed. In the 1980s, the phrase was used in the names of companies such as *Molecular Diagnostics Incorporated* and *Bethseda Research Laboratories Molecular Diagnostics.*

During the 1990s, the identification of newly discovered genes and new techniques for DNA sequencing led to the appearance of a distinct field of molecular and genomic laboratory medicine; in 1995, the *Association for Molecular Pathology* (AMP) was formed to give it structure. In 1999, the AMP co-founded *The Journal of Medical Diagnostics*. Informa Healthcare launched *Expert Reviews in Medical Diagnostics* in 2001. From 2002 onwards, the HapMap Project aggregated information on the one-letter genetic differences that recur in the human population—the single nucleotide polymorphisms—and their relationship with disease. In 2012, molecular diagnostic techniques for Thalassemia use genetic hybridization tests to identify the specific single nucleotide polymorphism causing an individual's disease.

As the commercial application of molecular diagnostics has become more important, so has the debate about patenting of the genetic discoveries at its heart. In 1998, the

European Union's Directive 98/44/ECclarified that patents on DNA sequences were allowable. In 2010 in the US, AMP sued Myriad Genetics to challenge the latter's patents regarding two genes, BRCA1, BRCA2, which are associated with breast cancer. In 2013, the U.S. Supreme Court partially agreed, ruling that a naturally occurring gene sequence could not be patented.

## Techniques

### Development from Research Tools

The industrialisation of molecular biology assay tools has made it practical to use them in clinics. Miniaturisation into a single handheld device can bring medical diagnostics into the clinic and into the office or home. The clinical laboratory requires high standards of reliability; diagnostics may require accreditation or fall under medical device regulations. As of 2011, some US clinical laboratories nevertheless used assays sold for "research use only".

Laboratory processes need to adhere to regulations, such as the Clinical Laboratory Improvement Amendments, Health Insurance Portability and Accountability Act, Good Laboratory Practice, and Food and Drug Administration specifications in the United States. Laboratory Information Management Systems help by tracking these processes. Regulation applies to both staff and supplies. As of 2012, twelve US states require molecular pathologists to be licensed; several boards such as the American Board of Medical Genetics and the American Board of Pathology certify technologists, supervisors, and laboratory directors.

Automation and sample barcoding maximise throughput and reduce the possibility of error or contamination during manual handling and results reporting. Single devices to do the assay from beginning to end are now available.

### Assays

Molecular diagnostics uses *in vitro* biological assays such as PCR-ELISA or Fluorescence in situ hybridization. The assay detects a molecule, often in low concentrations, that is a marker of disease or risk in a sample taken from a patient. Preservation of the sample before analysis is critical. Manual handling should be minimised. The fragile RNA molecule poses certain challenges. As part of the cellular process of expressing genes as proteins, it offers a measure of gene expression but it is vulnerable to hydrolysis and breakdown by ever-present RNAse enzymes. Samples can be snap-frozen in liquid nitrogen or incubated in preservation agents.

Because molecular diagnostics methods can detect sensitive markers, these tests are less intrusive than a traditional biopsy. For example, because cell-free nucleic acids exist in human plasma, a simple blood sample can be enough to sample genetic information from tumours, transplants or an unborn fetus. Many, but not all, molecular diagnostics methods based on nucleic acids detection use polymerase chain reaction (PCR) to vastly increase the number of nucleic acid molecules, thereby amplifying the target sequence(s) in the patient sample. PCR is a method that a template DNA is amplified using synthetic primers, a DNA polymerase, and dNTPs. The mixture is cycled between at least 2 temperatures: a high temperature for denaturing double-stranded DNA into single-stranded molecules and a low temperature for the primer to hybridize to the template and for the polymerase to extend the primer.

Each temperature cycle theoretically doubles the quantity of target sequence. Detection of sequence variations using PCR typically involves the design and use oligonucleotide reagents that amplify the variant of interest more efficiently than wildtype sequence. PCR is currently the most widely used method for detection of DNA sequences. The detection of the marker might use real time PCR, direct sequencing, microarray chips—prefabricated chips that test many markers at once, or MALDI-TOF The same principle applies to the proteome and the genome. High-throughput protein arrays can use complementary DNA or antibodies to bind and hence can detect many different proteins in parallel.

Molecular diagnostic tests vary widely in sensitivity, turn around time, cost, coverage and regulatory approval. They also vary in the level of validation applied in the laboratories using them. Hence, robust local validation in accordance with the regulatory requirements and use of appropriate controls is required especially where the result may be used to inform a patient treatment decision.

### Prenatal

Conventional prenatal tests for chromosomal abnormalities such as Down Syndrome rely on analysing the number and appearance of the chromosomes—the karyotype. Molecular diagnostics tests such as microarray comparative genomic hybridisation test a sample of DNA instead, and because of cell-free DNA in plasma, could be less invasive, but as of 2013 it is still an adjunct to the conventional tests.

### Treatment

Some of a patient's single nucleotide polymorphisms—slight differences in their DNA—can help predict how quickly they will metabolise particular drugs; this is called pharmacogenomics. For example, the enzyme CYP2C19 metabolises several drugs, such as the anti-clotting agent Clopidogrel, into their active forms. Some patients possess polymorphisms in specific places on the 2C19 gene that make poor metabolisers of those drugs; physicians can test for these polymorphisms and find out whether the drugs will be fully effective for that patient. Advances in molecular biology have helped show that some syndromes that were previously classed as a single disease are actually multiple subtypes with entirely different causes and treatments. Molecular diagnostics can help diagnose the subtype—for example of infections and cancers—or the genetic analysis of a disease with an inherited component, such as Silver-Russell syndrome.

### Infectious Disease

Molecular diagnostics are used to identify infectious diseases such as chlamydia, influenza virus and tuberculosis; or specific strains such as H1N1 virus or SARS-CoV-2. Genetic identification can be swift; for example a loop-mediated isothermal amplification test diagnoses the malaria parasite and is rugged enough for developing countries. But despite these advances in genome analysis, in 2013 infections are still more often identified by other means—their proteome, bacteriophage, or chromatographic profile. Molecular diagnostics are also used to understand the specific strain of the pathogen—for example by detecting which drug resistance genes it possesses—and hence which therapies to avoid. In addition, assays based on metagenomic next generation sequencing can be implemented to identify pathogenic organisms without bias.

### Disease Risk Management

A patient's genome may include an inherited or random mutation which affects the probability of developing a disease in the future. For example, Lynch syndrome is a genetic disease that predisposes patients to colorectal and other cancers; early detection can lead to close monitoring that improves the patient's chances of a good outcome. Cardiovascular risk is indicated by biological markers and screening can measure the risk that a child will be born with a genetic disease such as Cystic fibrosis. Genetic testing is ethically complex: patients may not want the stress of knowing their risk. In countries without universal healthcare, a known risk may raise insurance premiums.

### Cancer

Cancer is a change in the cellular processes that cause a tumour to grow out of control. Cancerous cells sometimes have mutations in oncogenes, such as KRAS and CTNNB1 (β-catenin). Analysing the molecular signature of cancerous cells—the DNA and its levels of expression via messenger RNA—enables physicians to characterise the cancer and to choose the best therapy for their patients. As of 2010, assays that incorporate an array of antibodies against specific protein marker molecules are an emerging technology; there are hopes for these multiplex assays that could measure many markers at once. Other potential future biomarkers include micro RNA molecules, which cancerous cells express more of than healthy ones.

Cancer is a disease with excessive molecular causes and constant evolution. There's also heterogeneity of disease even in an individual. Molecular studies of cancer have proved the significance of driver mutations in the growth and metastasis of tumors. Many technologies for detection of sequence variations have been developed for cancer research. These technologies generally can be grouped into three approaches: polymerase chain reaction (PCR), hybridization, and next-generation sequencing (NGS). Currently, a lot of PCR and hybridization assays have been approved by FDA as in vitro diagnostics. NGS assays, however, are still at an early stage in clinical diagnostics.

To do the molecular diagnostic test for cancer, one of the significant issue is the DNA sequence variation detection. Tumor biopsy samples used for diagnostics always contain as little as 5% of the target variant as compared to wildtype sequence. Also, for noninvasive applications from peripheral blood or urine, the DNA test must be specific enough to detect mutations at variant allele frequencies of less than 0.1%.

Currently, by optimizing the traditional PCR, there's a new invention, amplification-refractory mutation system (ARMS) is a method for detecting DNA sequence variants in cancer. The principle behind ARMS is that the enzymatic extension activity of DNA polymerases is highly sensitive to mismatches near the 3' end of primer. Many different companies have developed diagnostics tests based on ARMS PCR primers. For instance, Qiagen therascreen, Roche cobas and Biomerieux THxID have developed FDA approved PCR tests for detecting lung, colon cancer and metastatic melanoma mutations in the KRAS, EGFR and BRAF genes. Their IVD kits were basically validated on genomic DNA extracted from FFPE tissue.

There are also microarrays that utilize hybridization mechanism to diagnose cancer. More than a million of different probes can be synthesized on an array with Affymetrix's Genechip technology with a detection limit of one to ten copies of mRNA per well. Optimized microarrays are typically considered to produce repeatable relative quantitation of different targets. Currently, FDA have already approved a number of diagnostics assays utilizing microarrays: Agendia's MammaPrint assays can inform the breast cancer recurrence risk by profiling the expression of 70 genes related to breast cancer; Autogenomics INFNITI CYP2C19 assay can profile genetic polymorphisms, whose impacts on therapeutic response to antidepressants are great; and Affymetrix's CytoScan Dx can evaluate intellectual disabilities and congenital disorders by analyzing chromosomal mutation.

In the future, the diagnostic tools for cancer will likely to focus on the Next Generation Sequencing (NGS). By utilizing DNA and RNA sequencing to do cancer diagnostics, technology in the field of molecular diagnostics tools will develop better. Although NGS throughput and price have dramatically been reduced over the past 10 years by roughly 100-fold, we remain at least 6 orders of magnitude away from performing deep sequencing at a whole genome level. Currently, Ion Torrent developed some NGS panels based on translational AmpliSeq, for example, the Oncomine Comprehensive Assay. They are focusing on utilizing deep sequencing of cancer-related genes to detect rare sequence variants. Molecular diagnostics tool can be used for cancer risk assessment. For example, the BRCA1/2 test by Myriad Genetics assesses women for lifetime risk of breast cancer. Also, some cancers are not always employed with clear symptoms. It is useful to analyze people when they do not show obvious symptoms and thus can detect cancer at early stages. For example, the ColoGuard test may be used to screen people over 55 years old for colorectal cancer. Cancer is a longtime-scale disease with various progression steps, molecular diagnostics tools can be used for prognosis of cancer progression. For example, the OncoType Dx test by Genomic Health can estimate risk of breast cancer. Their technology can inform patients to seek chemotherapy when necessary by examining the RNA expression levels in breast cancer biopsy tissue.

With rising government support in DNA molecular diagnostics, it is expected that an increasing number of clinical DNA detection assays for cancers will become available soon. Currently, research in cancer diagnostics are developing fast with goals for lower cost, less time consumption and simpler methods for doctors and patients.

### 10.8.3 Data Analysis and Artificial Intelligence

Artificial intelligence (AI) and machine learning algorithms are being applied to interpret large datasets in clinical biochemistry. AI tools can help detect patterns in biochemical markers, predict disease risk, and recommend interventions, enhancing the accuracy and speed of diagnostics.

## 10.9 CONCLUSION

The principles of clinical biochemistry provide a framework for the accurate measurement and interpretation of biochemical markers that guide medical diagnosis and treatment. From analytical techniques to quality control and ethical considerations, these principles ensure the reliability of biochemical tests and the safety of patients. As clinical biochemistry continues to evolve with technological advancements, it will play an increasingly significant role in personalized medicine and healthcare innovation.

# CHAPTER 11
# MICROSCOPY

Microscopy is an invaluable technique in life sciences that enables the observation of structures too small to be seen by the naked eye. From investigating cellular components to studying complex tissue structures, microscopy allows scientists to explore the intricate details of biological specimens, advancing our understanding of life at the molecular and cellular levels.

## 11.1 BASIC PRINCIPLES OF MICROSCOPY

Microscopy is based on the principle of magnification, where a specimen is enlarged to reveal details not visible to the naked eye. This technique relies on lenses to bend light or electrons, increasing the size of the specimen's image. Another key principle is resolution, which is the ability to distinguish between two closely spaced points.

Microscopes use various types of illumination (such as light, fluorescence, or electron beams) and contrast methods to enhance visibility and distinguish different structures within a sample.

### 11.1.1 Magnification and Resolution

Magnification refers to the degree to which the image of a specimen is enlarged, while resolution is the microscope's ability to differentiate between two points that are close together. High magnification without adequate resolution results in a blurry image, underscoring the importance of both factors in microscopy. Resolution is influenced by factors such as the wavelength of light or electrons used and the quality of the lenses or electron optics in the microscope.

### 11.1.2 Contrast and Staining Techniques

In many specimens, structures are transparent and hard to differentiate from one another. Contrast enhancement and staining techniques help overcome this limitation. For light microscopy, stains bind selectively to various cellular components, such as nucleic acids or proteins, making them more visible. In fluorescence microscopy, fluorescent dyes or tags are used to label specific structures, providing high contrast and specificity. Electron microscopy uses heavy metal stains to enhance contrast, as electrons are scattered by the dense metal atoms, providing clearer images of ultrastructures.

## 11.2 TYPES OF MICROSCOPY

There are several types of microscopy, each suited for specific applications and providing unique insights into biological samples. Key microscopy types include light microscopy, electron microscopy, fluorescence microscopy, and atomic force microscopy.

### 11.2.1 Light Microscopy

Light microscopy, the most commonly used form, relies on visible light to illuminate the specimen. This type includes several subtypes, such as brightfield, darkfield, and phase-contrast microscopy.

- **Brightfield Microscopy**: This is the most basic type, where light passes through the specimen, with denser regions appearing darker. It is widely used for observing stained cells and tissues.
- **Darkfield Microscopy**: In darkfield microscopy, light is scattered by the specimen, making it appear bright against a dark background. This technique is useful for observing living, unstained specimens.
- **Phase-Contrast Microscopy**: Phase-contrast microscopy enhances contrast in transparent samples without staining, making it ideal for viewing live cells.

### 11.2.2 Fluorescence Microscopy

Fluorescence microscopy is a powerful technique in which fluorescent dyes or proteins are used to label specific components of the cell. When excited by a specific wavelength of light, these fluorophores emit light of a different wavelength, creating a bright, high-contrast image. This technique is invaluable for studying dynamic processes, protein localization, and cellular structures in both fixed and living cells. Advanced forms of fluorescence microscopy, such as confocal microscopy, allow for three-dimensional imaging of specimens.

### 11.2.3 Electron Microscopy

Electron microscopy (EM) uses electron beams instead of light, allowing for much higher resolution due to the shorter wavelength of electrons. This technique includes two main types:

- **Transmission Electron Microscopy (TEM)**: TEM involves passing electrons through an ultrathin specimen to visualize its internal structure. TEM offers high resolution, making it suitable for studying cellular ultrastructure, such as organelles and molecular complexes.
- **Scanning Electron Microscopy (SEM)**: SEM provides detailed surface images by scanning the specimen with an electron beam and detecting

the secondary electrons emitted. It is useful for studying surface morphology in detail and produces three-dimensional images of specimens.

### 11.2.4 Atomic Force Microscopy

Atomic force microscopy (AFM) is a type of scanning probe microscopy that measures the surface characteristics of a specimen using a sharp probe. AFM provides high-resolution, three-dimensional surface images and can be used on a variety of samples, including biological materials. Unlike electron microscopy, AFM can visualize living cells, as it does not require a vacuum environment, making it a valuable tool for studying cellular and molecular interactions.

## 11.3 APPLICATIONS OF MICROSCOPY IN LIFE SCIENCES

Microscopy plays a vital role across various fields of life sciences, enabling researchers to explore biological structures and processes at multiple levels. From basic research to medical diagnostics, microscopy provides insights essential for scientific advancement.

### 11.3.1 Cell Biology

Microscopy is foundational in cell biology, providing a window into the organization and function of cells. Techniques like fluorescence and electron microscopy allow scientists to observe cellular structures, organelles, and cytoskeletal networks. Time-lapse microscopy has further enabled the study of dynamic cellular processes, such as cell division, migration, and intracellular transport.

### 11.3.2 Molecular Biology

In molecular biology, microscopy techniques help visualize macromolecular interactions within the cellular environment. Fluorescence microscopy, combined with specific fluorescent labels, enables scientists to monitor protein localization, gene expression, and signal transduction pathways in real time. These insights are crucial for understanding gene function and regulation at the cellular level.

### 11.3.3 Histology and Pathology

Histology and pathology heavily rely on microscopy to study tissue organization and diagnose diseases. Microscopic examination of tissue samples reveals cellular abnormalities, inflammation, and tumors, aiding pathologists in diagnosing conditions such as cancer. Immunohistochemistry, a technique combining immunology and microscopy, is particularly useful in detecting specific antigens in tissues, helping to identify cancer markers or infection sites.

### 11.3.4 Microbiology

Microscopy is indispensable in microbiology for observing microorganisms such as bacteria, viruses, and fungi. Techniques like phase-contrast and darkfield microscopy allow researchers to study live microorganisms without staining, while fluorescence and electron microscopy are essential for identifying pathogens and studying microbial structures at high resolution.

## 11.4 ADVANCEMENTS IN MICROSCOPY

Recent advancements in microscopy have dramatically expanded its capabilities, pushing the boundaries of resolution, speed, and versatility. Innovations such as super-resolution microscopy, live-cell imaging, and cryo-electron microscopy have revolutionized the study of biological systems.

### 11.4.1 Super-Resolution Microscopy

Super-resolution microscopy techniques, such as STORM (stochastic optical reconstruction microscopy) and SIM (structured illumination microscopy), surpass the diffraction limit of conventional light microscopy. These methods enable imaging at nanometer resolution, making it possible to observe structures at the molecular level and revealing previously unseen details within cells.

### 11.4.2 Live-Cell Imaging

Live-cell imaging allows scientists to monitor cellular processes in real-time, providing insights into dynamic events such as protein trafficking, mitosis, and organelle movement. This technique has been instrumental in understanding cellular responses to environmental changes and drug treatments, offering a deeper understanding of cellular behavior under physiological conditions.

### 11.4.3 Cryo-Electron Microscopy

Cryo-electron microscopy (cryo-EM) has transformed structural biology by enabling the visualization of biomolecules at near-atomic resolution without the need for crystallization. Cryo-EM is used to study the structure of large protein complexes, viruses, and other biomolecules, offering valuable insights into their functions and mechanisms. The development of cryo-EM has been so impactful that it was awarded the Nobel Prize in Chemistry in 2017.

## 11.5 LIMITATIONS AND CHALLENGES IN MICROSCOPY

Despite its numerous applications, microscopy has certain limitations and challenges. For example, high-resolution techniques like electron microscopy require complex sample preparation and often damage living specimens, making it unsuitable for dynamic studies. Light microscopy, while versatile, is limited by the diffraction limit, which restricts resolution to around 200 nanometers. Additionally, microscopy can be expensive, and the complexity of advanced techniques requires specialized training for accurate data acquisition and interpretation.

## 11.6 FUTURE DIRECTIONS IN MICROSCOPY

The future of microscopy lies in continued innovation, particularly in improving resolution, sensitivity, and speed. Hybrid approaches, such as correlative light and electron microscopy (CLEM), are being developed to combine the advantages of different microscopy techniques, providing comprehensive information on both cellular structure and function. Moreover, advancements in automation and artificial intelligence are expected to streamline image analysis, making it faster and more efficient. The integration of artificial intelligence in microscopy may also enhance diagnostic accuracy, aiding in the detection of subtle abnormalities and accelerating research progress.

## 11.7 CONCLUSION

Microscopy is a cornerstone of life sciences, providing essential insights into biological structures and processes at the microscopic and molecular levels. From the basic principles of magnification and resolution to advanced applications in fields like cell biology, pathology, and microbiology, microscopy has greatly enhanced our understanding of the natural world. Technological advancements continue to push the capabilities of microscopy, promising even greater potential for discovery. As microscopy evolves, it will undoubtedly remain an indispensable tool for researchers, shaping future breakthroughs in science and medicine.

# CHAPTER 12

# CENTRIFUGATION AND ULTRACENTRIFUGATION

Centrifugation and ultracentrifugation are powerful techniques extensively used in biological and chemical research for the separation of particles based on their size, shape, and density. These methods are fundamental in various applications, including the isolation of cellular components, purification of proteins and nucleic acids, and characterization of macromolecules. By employing centrifugal force, these techniques facilitate the analysis and manipulation of biological samples, providing researchers with crucial insights into cellular processes, molecular interactions, and biochemical pathways.

## 12.1 PRINCIPLES OF CENTRIFUGATION

Centrifugation is based on the principle of sedimentation, where particles in a solution experience a force that causes them to move and settle. When a sample is spun in a centrifuge, the centrifugal force acts on the particles, causing them to migrate outward from the center of rotation. The sedimentation process is influenced by various factors, including the particle's size, density, shape, and the viscosity of the surrounding medium.

### 12.1.1 Sedimentation Rate

The sedimentation rate of particles is determined by their mass and the applied centrifugal force. Larger and denser particles experience greater forces and sediment more quickly than smaller or less dense particles. The sedimentation process can be described mathematically using the sedimentation coefficient, which relates to the velocity of the particle's movement in a gravitational or centrifugal field.

### 12.1.2 Relative Centrifugal Force (RCF)

Relative centrifugal force (RCF) quantifies the centrifugal force experienced by the sample and is expressed in terms of the gravitational force (g). RCF is calculated based on the speed of rotation (in revolutions per minute, RPM) and the radius of the centrifuge rotor. Ultracentrifuges can generate extremely high RCF values, making them suitable for the separation of very small particles, such as organelles, macromolecules, and viruses.

## 12.2 TYPES OF CENTRIFUGATION

Centrifugation can be categorized into different types, each tailored for specific applications. The two primary categories are preparative centrifugation and analytical centrifugation.

### 12.2.1 Preparative Centrifugation

Preparative centrifugation is aimed at isolating specific components from a mixture. This technique is vital in cell biology, biochemistry, and molecular biology, enabling the purification of proteins, nucleic acids, and organelles.

#### Differential Centrifugation

In biochemistry and cell biology, differential centrifugation (also known as differential velocity centrifugation) is a common procedure used to separate organelles and other sub-cellular particles based on their sedimentation rate. Although often applied in biological analysis, differential centrifugation is a general technique also suitable for crude purification of non-living suspended particles (e.g. nanoparticles, colloidal particles, viruses).

In a typical case where differential centrifugation is used to analyze cell-biological phenomena (e.g. organelle distribution), a tissue sample is first lysed to break the cell membranes and release the organelles and cytosol. The lysate is then subjected to repeated centrifugations, where particles that sediment sufficiently quickly at a given centrifugal force for a given time form a compact "pellet" at the bottom of the centrifugation tube.

After each centrifugation, the *supernatant* (non-pelleted solution) is removed from the tube and re-centrifuged at an increased centrifugal force and/or time. Differential centrifugation is suitable for crude separations on the basis of sedimentation rate, but more fine grained purifications may be done on the basis of density through equilibrium density-gradient centrifugation.

Thus, the differential centrifugation method is the successive pelleting of particles from the previous supernatant, using increasingly higher centrifugation forces. Cellular organelles separated by differential centrifugation maintain a relatively high degree of normal functioning, as long as they are not subject to denaturing conditions during isolation.

#### Theory

In a viscous fluid, the *rate* of sedimentation of a given suspended particle (as long as the particle is denser than the fluid) is largely a function of the following factors:

- Gravitational force.
- Difference in density.

- Fluid viscosity.
- Particle size and shape.

Larger particles sediment more quickly and at lower centrifugal forces. If a particle is less dense than the fluid (e.g., fats in water), the particle will not sediment, but rather will float, regardless of strength of the g-force experienced by the particle. Centrifugal force separates components not only on the basis of density, but also of particle size and shape. In contrast, a more specialized equilibrium density-gradient centrifugation produces a separation profile dependent on particle-density alone, and therefore is suitable for more fine-grained separations.

High g-force makes sedimentation of small particles much faster than Brownian diffusion, even for very small (nanoscale) particles. When a centrifuge is used, Stokes' law must be modified to account for the variation in g-force with distance from the center of rotation.

$$D = \sqrt{\frac{18\eta \ln(R_f / R_i)}{(\rho_p - \rho_f)\omega^2 t}}$$

where

- D is the minimum diameter of the particles expected to sediment (m)
- $\eta$ (or $\mu$) is the fluid dynamic viscosity (Pa.s)
- $R_f$ is the final radius of rotation (m)
- $R_i$ is the initial radius of rotation (m)
- $\rho_p$ is particle volumetric mass density (kg/m$^3$)
- $\rho_f$ is the fluid volumetric mass density (kg/m$^3$)
- $\omega$ is the angular velocity (radian/s)
- t is the time required to sediment from $R_i$ to $R_f$ (s)

## Ultracentrifugation

The lysed sample is now ready for centrifugation in an ultracentrifuge. An ultracentrifuge consists of a refrigerated, low-pressure chamber containing a rotor which is driven by an electrical motor capable of high speed rotation. Samples are placed in tubes within or attached to the rotor. Rotational speed may reach up to 100,000 rpm for floor model, 150,000 rpm for bench-top model (Beckman Optima Max-XP or Sorvall MTX150 or himac CS150NX), creating centrifugal speed forces of 800,000g to 1,000,000g. This force causes sedimentation of macromolecules, and can even cause non-uniform distributions of small molecules.

Since different fragments of a cell have different sizes and densities, each fragment will settle into a pellet with different minimum centrifugal forces. Thus, separation of the sample into different layers can be done by first centrifuging the original lysate under weak forces, removing the pellet, then exposing the subsequent supernatants to sequentially greater centrifugal fields. Each time a portion of different density is sedimented to the bottom of the container and extracted, and repeated application produces a rank of layers which includes different parts of the original sample. Additional steps can be taken to further refine each of the obtained pellets.

Sedimentation depends on mass, shape, and partial specific volume of a macromolecule, as well as solvent density, rotor size and rate of rotation. The sedimentation velocity can be monitored during the experiment to calculate molecular weight. Values of sedimentation coefficient (S) can be calculated. Large values of S (faster sedimentation rate) correspond to larger molecular weight. Dense particle sediments more rapidly. Elongated proteins have larger frictional coefficients, and sediment more slowly to ensure accuracy.

#### Differences between Differential and Density Gradient Centrifugation

The difference between differential and density gradient centrifugation techniques is that the latter method uses solutions of different densities (e.g. sucrose, Ficoll, Percoll) or gels through which the sample passes. This separates the sample into layers by relative density, based on the principle that molecules settle down under a centrifugal force until they reach a medium with the density the same as theirs. The degree of separation or number of layers depends on the solution or gel. Differential centrifugation, on the other hand, does not utilize a density gradient, and the centrifugation is taken in increasing speeds. The different centrifugation speeds often create separation into not more than two fractions, so the supernatant can be separated further in additional centrifugation steps. For that, each step the centrifugation speed has to be increased until the desired particles are separated. In contrast, the density gradient centrifugation is usually performed with just one centrifugation speed.

### Density Gradient Centrifugation

Density gradient centrifugation is another powerful method used to separate particles based on their buoyant density. In this technique, a gradient of a dense medium, such as sucrose or cesium chloride, is created in a centrifuge tube. When the sample is centrifuged, particles will migrate to the point in the gradient where their density matches that of the surrounding medium. This method is particularly effective for purifying subcellular organelles, macromolecules, and viruses.

## 12.2.2 Analytical Centrifugation

Analytical centrifugation is used primarily for studying the properties of macromolecules in solution rather than isolating them. This method allows researchers to analyze the sedimentation behavior of molecules and obtain information about their size, shape, and molecular weight.

## Analytical Ultracentrifugation

Analytical ultracentrifugation is an analytical technique which combines an ultracentrifuge with optical monitoring systems. In an analytical ultracentrifuge (commonly abbreviated as AUC), a sample's sedimentation profile is monitored in real time by an optical detection system. The sample is detected via ultraviolet light absorption and/or interference optical refractive index sensitive system, monitored by light-sensitive diode array or by film in the older machines. The operator can thus observe the change of sample concentration versus the axis of the rotation profile with time as a result of the applied centrifugal field. With modern instrumentation, these observations are electronically digitized and stored for further mathematical analysis.

The information that can be obtained from an analytical ultracentrifuge includes the gross shape of macromolecules, conformational changes in macromolecules, and size distributions of macromolecules. With AUC it is possible to gain information on the number and subunit stoichiometry of non-covalent complexes and equilibrium constants of macromolecules such as proteins, DNA, nanoparticles or other assemblies from different molecule classes. The simplest measurement to be obtained is the sedimentation coefficient, which depends upon the size of the molecules being sedimented. This is the ratio of a particle's sedimentation velocity to the applied acceleration causing the sedimentation.

Analytical ultracentrifugation has recently seen a rise in use because of increased ease of analysis with modern computers and the development of software, including a National Institutes of Health supported software package, SedFit.

An analytical ultracentrifuge has a light source and optical detectors. To allow the light to pass through the analyte during the ultracentrifuge run, specialized cells are required which have to meet high optical standards as well as to resist the centrifugal forces. Each cell consists of a housing, two windows made from optically pure quartz glass, and a centrepiece with one or two sectors and filling holes for the sector(s), closed with a screw plug in the housing. These cell are placed into a rotor cavity with a continuous bore, with a collar at the bottom to retain the cell.

### Sedimentation Velocity

Sedimentation velocity experiments render the shape and molar mass of the analytes, as well as their size-distribution. The size resolution of this method scales approximately with the square of the particle radii, and by adjusting the rotor speed of the experiment size-ranges from 100 Da to 10 GDa can be covered. Sedimentation velocity experiments can also be used to study reversible chemical equilibria between macromolecular species, by either monitoring the number and molar mass of macromolecular complexes, by gaining information about the complex composition from multi-signal analysis exploiting differences in each components spectroscopic signal, or by following the composition dependence of the sedimentation rates of the macromolecular system, as described in Gilbert-Jenkins theory. The experiment aims to monitor the sedimentation behavior at a fixed angular speed.

#### Sedimentation Equilibrium

Sedimentation equilibrium experiments reports the molar mass of analytes and their chemical equilibrium constants. The rotor speed is adjusted such that a steady-state concentration profile c(r) of the sample in the cell is formed, where sedimentation and diffusion cancel out each other.

## 12.3 ULTRACENTRIFUGATION: AN ADVANCED TECHNIQUE

Ultracentrifugation refers to a specialized form of centrifugation that employs extremely high centrifugal forces to separate small particles, such as proteins, nucleic acids, and viruses. Ultracentrifuges operate at speeds that can exceed 100,000 RPM, generating RCF values in the range of millions of g.

### 12.3.1 Principles of Ultracentrifugation

Ultracentrifugation is based on the same sedimentation principles as conventional centrifugation, but with several key differences. The high speeds achieved in ultracentrifugation allow for the separation of particles that are too small to be effectively isolated by standard centrifugation methods. The process typically involves running samples in a vacuum to minimize heat generation and prevent sample degradation.

### 12.3.2 Types of Ultracentrifugation

Ultracentrifugation can be performed in both preparative and analytical modes, depending on the research needs.

#### Preparative Ultracentrifugation

Preparative ultracentrifugation is used to isolate and purify specific biological components. It is particularly effective for separating subcellular organelles, large protein complexes, and viral particles. By employing density gradients, researchers can achieve high-resolution separations, allowing for the collection of pure fractions for subsequent analysis.

#### Analytical Ultracentrifugation

Analytical ultracentrifugation is primarily used for studying the physical and chemical properties of macromolecules. By monitoring the sedimentation process, researchers can gain insights into the molecular weight, shape, and interactions of proteins and nucleic acids. This technique is particularly valuable in understanding protein folding, complex formation, and molecular interactions.

## 12.4 APPLICATIONS OF CENTRIFUGATION AND ULTRACENTRIFUGATION

Centrifugation and ultracentrifugation have a wide range of applications across various scientific disciplines. Their ability to separate and isolate biological components is crucial for advancing research in fields such as cell biology, biochemistry, and molecular biology.

### 12.4.1 Cell Fractionation

Cell fractionation is one of the primary applications of centrifugation techniques. By breaking down cells and separating their components, researchers can study the functions and interactions of organelles. Differential centrifugation allows for the isolation of nuclei, mitochondria, and other organelles, providing insights into their roles in cellular processes.

### 12.4.2 Protein Purification

Centrifugation is a key step in the purification of proteins from complex biological mixtures. Techniques such as density gradient ultracentrifugation enable the separation of proteins based on their size and density, allowing researchers to isolate specific proteins for functional studies or structural analysis.

### 12.4.3 Isolation of Nucleic Acids

Centrifugation methods, particularly density gradient centrifugation, are widely used for isolating nucleic acids, such as DNA and RNA. By separating nucleic acids based on their density, researchers can obtain high-quality samples for various applications, including cloning, sequencing, and gene expression analysis.

### 12.4.4 Viral Studies

Ultracentrifugation plays a crucial role in virology, where it is used to isolate and purify viruses from infected cell cultures or tissue samples. The ability to separate viral particles from cellular debris allows for the study of virus structure, infectivity, and life cycles, contributing to our understanding of viral diseases and vaccine development.

## 12.5 ADVANTAGES AND LIMITATIONS

Centrifugation and ultracentrifugation offer several advantages in laboratory settings. They are relatively quick, reproducible, and allow for the separation of complex mixtures. Additionally, these techniques are versatile and can be adapted for various applications, making them essential tools in scientific research. However, there are also limitations to these methods. The need for specialized equipment, such as high-speed centrifuges and ultracentrifuges, can make them costly and less accessible for smaller laboratories. Moreover, the handling of samples during centrifugation requires careful attention to avoid contamination or loss of material.

## 12.6 FUTURE DIRECTIONS IN CENTRIFUGATION AND ULTRACENTRIFUGATION

As research continues to evolve, the future of centrifugation and ultracentrifugation is likely to see advancements in several areas. Improved technology will enhance speed, efficiency, and resolution, enabling the separation of even smaller particles. Additionally, automation and integration of centrifugation with other analytical techniques may lead to more streamlined workflows in laboratories.

## 12.7 CONCLUSION

Centrifugation and ultracentrifugation are indispensable techniques in biological research, allowing for the effective separation and analysis of various cellular components, proteins, and nucleic acids. By harnessing the principles of sedimentation, these methods have become essential tools in advancing our understanding of biological systems and processes. With ongoing technological advancements, centrifugation and ultracentrifugation will continue to play a pivotal role in scientific discovery and innovation.

# CHAPTER 13
# SPECTROSCOPIC TECHNIQUES

Spectroscopic techniques are powerful analytical methods widely used across various scientific disciplines, including chemistry, biology, physics, and materials science. These techniques rely on the interaction of electromagnetic radiation with matter to obtain detailed information about the structure, composition, and properties of substances. Spectroscopy provides qualitative and quantitative data, making it an invaluable tool for researchers and scientists in their quest to understand the molecular and atomic properties of materials.

## 13.1 PRINCIPLES OF SPECTROSCOPY

At the core of spectroscopic techniques is the principle that matter interacts with electromagnetic radiation in specific ways, depending on the type of radiation and the properties of the material being analyzed. When electromagnetic radiation, which encompasses a broad range of wavelengths from gamma rays to radio waves, interacts with matter, it can be absorbed, emitted, or scattered. These interactions provide insights into the energy levels of molecules and atoms, allowing scientists to infer structural information and identify compounds.

### 13.1.1 Types of Electromagnetic Radiation

Electromagnetic radiation can be categorized based on its wavelength or frequency. The spectrum is divided into several regions, including ultraviolet (UV), visible, infrared (IR), microwave, and radio frequencies. Each region interacts differently with matter, leading to specific spectroscopic techniques that target different types of molecular transitions.

- **Ultraviolet and Visible Spectroscopy (UV-Vis):** This technique involves the absorption of UV or visible light, causing electronic transitions in molecules. It is commonly used for analyzing conjugated systems, transition metal complexes, and biological molecules such as proteins and nucleic acids.
- **Infrared Spectroscopy (IR):** Infrared radiation is absorbed by molecules, causing vibrational transitions. This technique is particularly useful for identifying functional groups and studying molecular vibrations.
- **Nuclear Magnetic Resonance Spectroscopy (NMR):** NMR relies on

the interaction of atomic nuclei with radiofrequency radiation in a strong magnetic field. This technique provides detailed information about the local environment of specific nuclei, making it invaluable in determining molecular structure and dynamics.

- **Mass Spectrometry (MS):** Although not a spectroscopic technique in the traditional sense, mass spectrometry is often combined with spectroscopic methods. It measures the mass-to-charge ratio of ions generated from a sample, providing insights into molecular weight and composition.

## 13.2 TYPES OF SPECTROSCOPIC TECHNIQUES

### 13.2.1 Ultraviolet-Visible Spectroscopy (UV-Vis)

UV-Vis spectroscopy involves measuring the absorbance of UV and visible light by a sample. The technique is based on the Beer-Lambert law, which relates the absorbance of light to the concentration of the absorbing species and the path length of the light through the sample. UV-Vis spectroscopy is widely used in quantitative analysis, particularly for determining concentrations of colored compounds, assessing the purity of substances, and studying electronic transitions in molecules.

### 13.2.2 Infrared Spectroscopy (IR)

Infrared spectroscopy measures the absorption of infrared radiation by a sample, leading to molecular vibrations. Each bond in a molecule has a characteristic vibrational frequency, allowing researchers to identify functional groups and molecular structures. IR spectroscopy is commonly used in organic chemistry for identifying compounds, characterizing polymers, and analyzing functional groups in biomolecules. Fourier-transform infrared spectroscopy (FTIR) is a modern variant that enhances sensitivity and speed.

### 13.2.3 Nuclear Magnetic Resonance Spectroscopy (NMR)

NMR spectroscopy is a non-destructive technique that provides detailed information about the structure and dynamics of molecules. It exploits the magnetic properties of certain nuclei, typically hydrogen (^1H) and carbon (^13C), in the presence of a strong magnetic field. NMR spectra reveal information about the number of nuclei, their chemical environment, and their connectivity, making it essential for elucidating molecular structures in organic and medicinal chemistry.

### 13.2.4 Mass Spectrometry (MS)

Mass spectrometry is a technique that ionizes chemical species and sorts the ions based on their mass-to-charge ratio. The resulting mass spectrum provides information about the molecular weight, structure, and abundance of different ions. MS is often coupled with chromatographic techniques for compound separation before analysis, making it a powerful tool for analyzing complex mixtures, including biological samples, environmental pollutants, and synthetic compounds.

### 13.2.5 Fluorescence Spectroscopy

Fluorescence spectroscopy involves the excitation of a sample with a specific wavelength of light, leading to the emission of light at a longer wavelength. This technique is highly sensitive and is often used in biological applications, such as studying proteins, nucleic acids, and cellular processes. Fluorescence spectroscopy can provide information about molecular interactions, conformational changes, and localization within cells.

### 13.2.6 Raman Spectroscopy

Raman spectroscopy is based on inelastic scattering of monochromatic light (usually from a laser) by molecules. When light interacts with a sample, it can scatter with a change in energy, providing information about vibrational, rotational, and other low-frequency modes in the system. Raman spectroscopy is particularly useful for analyzing chemical bonds and molecular symmetry and is often used in the study of solids, liquids, and gases.

## 13.3 APPLICATIONS OF SPECTROSCOPIC TECHNIQUES

Spectroscopic techniques have a wide range of applications across various fields, including chemistry, biology, medicine, and materials science.

### 13.3.1 Chemical Analysis

Spectroscopy is extensively used for qualitative and quantitative analysis of chemical compounds.

Researchers use UV-Vis, IR, and NMR spectroscopy to identify functional groups, determine molecular structures, and analyze mixtures. Mass spectrometry is often employed to confirm molecular weights and elucidate complex structures.

### 13.3.2 Biological and Medical Research

In biological research, spectroscopic techniques are crucial for studying biomolecules, cellular processes, and metabolic pathways. UV-Vis and fluorescence spectroscopy are widely used for analyzing proteins, nucleic acids, and enzyme kinetics.

NMR spectroscopy plays a vital role in drug discovery, enabling the identification of ligand-receptor interactions and conformational changes in biological macromolecules.

### 13.3.3 Environmental Monitoring

Spectroscopic techniques are employed in environmental monitoring to detect pollutants and assess the quality of air, water, and soil. UV-Vis and fluorescence spectroscopy can measure the concentration of organic compounds in environmental samples, while mass spectrometry is used to identify trace contaminants in complex mixtures.

### 13.3.4 Material Characterization

Spectroscopy is essential in materials science for characterizing polymers, nanomaterials, and other solid-state materials. Techniques such as IR and Raman spectroscopy provide insights into molecular structure and bonding, while NMR spectroscopy can help understand the properties of materials at the molecular level.

## 13.4 ADVANTAGES AND LIMITATIONS OF SPECTROSCOPIC TECHNIQUES

Spectroscopic techniques offer several advantages that make them indispensable in scientific research. They are often non-destructive, allowing for the analysis of precious samples without altering their properties. Many spectroscopic methods are highly sensitive, enabling the detection of low concentrations of analytes. Additionally, they can provide rapid results and be used in real-time analysis.

However, there are also limitations to consider. Some techniques may require complex sample preparation, and the interpretation of spectra can be challenging due to overlapping signals or matrix effects. Furthermore, certain methods may have limited applicability for specific types of samples or analytes.

## 13.5 FUTURE DIRECTIONS IN SPECTROSCOPIC TECHNIQUES

As technology advances, the future of spectroscopic techniques is poised for significant developments. Improvements in instrumentation, such as miniaturization and enhanced sensitivity, will facilitate the analysis of smaller samples and lower concentrations. Moreover, the integration of machine learning and artificial intelligence will enhance data analysis, allowing for more accurate interpretation of complex spectra.

## 13.6 CONCLUSION

Spectroscopic techniques are invaluable tools in modern science, providing critical insights into the composition, structure, and behavior of various materials. From chemical analysis to biological research and environmental monitoring, spectroscopy has transformed our understanding of molecular and atomic interactions. As advancements continue to unfold, these techniques will undoubtedly play a pivotal role in driving innovation and discovery across diverse fields of study.

# CHAPTER 14

# BASIC TECHNIQUES PROBING MOLECULAR STRUCTURE AND INTERACTIONS

Understanding molecular structure and interactions is essential in fields such as chemistry, biology, and materials science. Various techniques have been developed to probe these aspects, each offering unique insights into the characteristics of molecules.

## 14.1 X-RAY CRYSTALLOGRAPHY

### 14.1.1 Principle and Technique

X-ray crystallography is one of the most powerful techniques for determining the three-dimensional structure of molecules at atomic resolution. The technique relies on the diffraction of X-rays by the electron cloud of a crystalline sample. When X-rays are directed at a crystal, they interact with the electrons and produce a diffraction pattern. By analyzing this pattern using mathematical methods, researchers can reconstruct the electron density map, revealing the arrangement of atoms within the molecule.

### 14.1.2 Applications

X-ray crystallography has been pivotal in various scientific fields, including drug design, biochemistry, and materials science. It has enabled the determination of the structures of complex biomolecules such as proteins, nucleic acids, and carbohydrates, providing insights into their functions and interactions. Notably, the elucidation of the double helix structure of DNA by James Watson and Francis Crick relied heavily on X-ray diffraction data.

## 14.2 NUCLEAR MAGNETIC RESONANCE (NMR) SPECTROSCOPY

### 14.2.1 Principle and Technique

Nuclear Magnetic Resonance (NMR) spectroscopy is a non-destructive technique that exploits the magnetic properties of certain atomic nuclei, most commonly hydrogen (^1H) and carbon (^13C). When placed in a strong magnetic field, these nuclei resonate at specific frequencies when exposed to radiofrequency radiation.

The resulting NMR spectrum provides information about the chemical environment of the nuclei, allowing researchers to deduce molecular structure, dynamics, and interactions.

### 14.2.2 Applications

NMR spectroscopy is widely used in organic chemistry, biochemistry, and medicinal chemistry for structural elucidation and dynamics studies.

It is particularly valuable for analyzing small molecules, proteins, and nucleic acids in solution. Researchers can obtain detailed information about molecular conformations, dynamic processes, and ligand-receptor interactions, making NMR an essential tool in drug discovery and development.

## 14.3 MASS SPECTROMETRY (MS)

### 14.3.1 Principle and Technique

Mass spectrometry (MS) is an analytical technique used to determine the mass-to-charge ratio of ions. The basic principle involves ionizing chemical compounds and then sorting the ions based on their mass.

The resulting mass spectrum provides information about the molecular weight, structure, and abundance of different species in a sample. Techniques such as electrospray ionization (ESI) and matrix-assisted laser desorption/ionization (MALDI) have significantly advanced the capabilities of MS.

### 14.3.2 Applications

Mass spectrometry is extensively used in various fields, including proteomics, metabolomics, and environmental analysis. It enables the identification and quantification of complex mixtures, such as proteins in biological samples or contaminants in environmental samples.

Moreover, when coupled with chromatographic techniques, MS provides a robust platform for analyzing diverse compounds, from small metabolites to large biomolecules.

## 14.4 INFRARED (IR) SPECTROSCOPY

### 14.4.1 Principle and Technique

Infrared (IR) spectroscopy is based on the absorption of infrared radiation by molecules, leading to vibrational transitions. Each molecular bond has a characteristic vibrational frequency, allowing researchers to identify functional groups and molecular structures. The absorption spectrum generated provides information about the types of bonds present in a molecule, revealing functional group characteristics.

### 14.4.2 Applications

IR spectroscopy is commonly used in organic and inorganic chemistry for identifying functional groups and characterizing compounds. It plays a significant role in the analysis of polymers, pharmaceuticals, and natural products.

In addition, FTIR (Fourier-transform infrared) spectroscopy, a modern variant, enhances sensitivity and speed, making it suitable for various applications, including quality control and forensic analysis.

## 14.5 ULTRAVIOLET-VISIBLE (UV-VIS) SPECTROSCOPY

### 14.5.1 Principle and Technique

Ultraviolet-visible (UV-Vis) spectroscopy involves measuring the absorption of UV and visible light by a sample. The absorption of light results in electronic transitions within the molecule, which can be correlated to its structure and concentration.

The technique follows Beer-Lambert's law, which states that absorbance is proportional to the concentration of the absorbing species and the path length of light through the sample

### 14.5.2 Applications

UV-Vis spectroscopy is widely used in quantitative analysis and is particularly effective for analyzing colored compounds and conjugated systems. It is commonly employed in pharmaceutical analysis, environmental monitoring, and biological studies to determine the concentration of specific analytes in solution.

Additionally, UV-Vis spectroscopy can provide information about molecular interactions, such as binding events between ligands and receptors.

## 14.6 CIRCULAR DICHROISM (CD) SPECTROSCOPY

### 14.6.1 Principle and Technique

Circular dichroism (CD) spectroscopy is a chiroptical technique that measures the differential absorption of left- and right-handed circularly polarized light by optically active molecules.

This property arises from the molecular asymmetry and provides insights into the secondary structure of proteins and other biomolecules. The resulting CD spectrum reflects the conformation of the chiral molecules, enabling structural analysis.

### 14.6.2 Applications

CD spectroscopy is widely used in biochemistry and molecular biology to study the secondary structure of proteins and nucleic acids. It is a valuable tool for monitoring conformational changes in proteins upon ligand binding or changes in environmental conditions. CD spectroscopy is often used in conjunction with other techniques, such as NMR and X-ray crystallography, to provide complementary structural information.

## 14.7 SURFACE PLASMON RESONANCE (SPR)

### 14.7.1 Principle and Technique

Surface plasmon resonance (SPR) is a sensitive technique used to study molecular interactions in real-time.

It relies on the excitation of surface plasmons—coherent electron oscillations at the interface between a metal and a dielectric material—when light hits the surface at a specific angle.

Changes in the refractive index near the sensor surface, caused by the binding of analytes, result in shifts in the resonance angle, which can be measured.

### 14.7.2 Applications

SPR is widely used in biomolecular interaction studies, particularly for characterizing protein-protein, protein-DNA, and protein-ligand interactions. It enables real-time monitoring of binding events, providing information about affinity, kinetics, and specificity. This technique is crucial in drug discovery and development, allowing researchers to assess the effectiveness of potential therapeutic agents.

## 14.8 ATOMIC FORCE MICROSCOPY (AFM)

### 14.8.1 Principle and Technique

Atomic force microscopy (AFM) is a type of scanning probe microscopy that provides high-resolution images of surfaces at the nanoscale. AFM employs a cantilever with a sharp tip that scans the surface of a sample. The interaction forces between the tip and the surface are measured, providing topographical information and mechanical properties at the molecular level.

### 14.8.2 Applications

AFM is widely used in materials science, biology, and nanotechnology for characterizing surfaces and studying molecular interactions. It can provide insights into the morphology of biomolecules, the mechanical properties of materials, and the interactions between proteins and surfaces. AFM's ability to operate in various environments, including liquid and air, makes it suitable for studying biological samples in their native states.

## 14.9 CONCLUSION

The study of molecular structure and interactions is fundamental to advancing our understanding of chemistry, biology, and materials science. A range of techniques, including X-ray crystallography, NMR spectroscopy, mass spectrometry, and various spectroscopic methods, provides researchers with invaluable tools to probe these complex systems.

Each technique offers unique insights and applications, from elucidating molecular structures to characterizing dynamic interactions. As advancements in technology continue to evolve, these techniques will undoubtedly play a crucial role in driving scientific discovery and innovation in the years to come.

# CHAPTER 15
# MASS SPECTROMETRIC TECHNIQUES

Mass spectrometry (MS) is an analytical technique that measures the mass-to-charge ratio of ions. It has become an indispensable tool in various fields, including chemistry, biochemistry, and environmental science. The versatility of mass spectrometric techniques allows for the analysis of complex mixtures, the determination of molecular weights, and the identification of chemical structures.

## 15.1 PRINCIPLES OF MASS SPECTROMETRY

### 15.1.1 Basic Components of Mass Spectrometry

Mass spectrometry involves three primary stages: ionization, acceleration, and detection. The process begins with ionization, where molecules are converted into ions. This can occur through various methods, such as electron impact (EI), chemical ionization (CI), and electrospray ionization (ESI). Once ionized, the ions are accelerated by an electric field, which propels them into a mass analyzer. The mass analyzer separates the ions based on their mass-to-charge ratio (m/z). Finally, the ions are detected, producing a mass spectrum that provides information about the composition and structure of the analytes.

### 15.1.2 The Mass Spectrum

The resulting mass spectrum is a graphical representation of the detected ions, with the x-axis representing the mass-to-charge ratio (m/z) and the y-axis representing the relative abundance of each ion. The peaks in the mass spectrum correspond to different ions present in the sample, allowing researchers to identify and quantify the components of complex mixtures. The interpretation of mass spectra involves understanding the fragmentation patterns of ions, which can provide insights into the structural characteristics of the molecules.

## 15.2 IONIZATION TECHNIQUES

### 15.2.1 Electron Impact Ionization (EI)

Electron impact ionization is one of the oldest and most widely used ionization methods in mass spectrometry. In EI, high-energy electrons are directed at the analyte gas, causing the ejection of electrons from the molecules and resulting in the formation of positive ions. EI is particularly effective for small, volatile organic compounds and provides fragmentation patterns that are useful for structural elucidation. However, it is less suitable for large biomolecules due to the extensive fragmentation that occurs.

### 15.2.2 Chemical Ionization (CI)

Chemical ionization is a softer ionization technique compared to EI. In CI, a reagent gas is introduced into the ionization chamber, where it interacts with the analyte to produce ions. This method often results in less fragmentation, making it easier to determine the molecular weight of the analyte. CI is particularly useful for compounds that are difficult to ionize using EI and is commonly employed in conjunction with EI to enhance structural information.

### 15.2.3 Electrospray Ionization (ESI)

Electrospray ionization is a powerful technique for ionizing large biomolecules, such as proteins and nucleic acids. In ESI, a high-voltage electric field is applied to a solution containing the analyte, causing the formation of charged droplets. As the solvent evaporates, charged ions are released into the gas phase. ESI is compatible with liquid chromatography, making it an excellent choice for analyzing complex biological samples. This technique allows for the determination of molecular weights and the study of protein-ligand interactions in real time.

### 15.2.4 Matrix-Assisted Laser Desorption/Ionization (MALDI)

MALDI is another soft ionization technique that is particularly suited for the analysis of large biomolecules. In this method, the analyte is mixed with a matrix material that absorbs laser energy. Upon laser irradiation, the matrix promotes the desorption and ionization of the analyte molecules. MALDI is commonly used for the analysis of proteins, peptides, and oligonucleotides. The technique offers high sensitivity and is compatible with various separation techniques, including gel electrophoresis.

## 15.3 MASS ANALYZERS

### 15.3.1 Quadrupole Mass Analyzer

The quadrupole mass analyzer consists of four parallel rods that create an oscillating electric field. Ions are filtered based on their stability within the oscillating field, allowing only ions of a specific m/z ratio to pass through to the detector. Quadrupoles are commonly used in tandem mass spectrometry (MS/MS) experiments, where ions are fragmented for structural analysis. This analyzer is versatile and widely used in various applications, from environmental analysis to proteomics.

### 15.3.2 Time-of-Flight (TOF) Mass Analyzer

Time-of-flight mass analyzers measure the time it takes for ions to travel a fixed distance. When ions are generated, they are accelerated by an electric field and allowed to drift through a vacuum. The time taken by each ion to reach the detector is inversely proportional to its mass, allowing for the determination of m/z ratios. TOF analyzers offer high resolution and the ability to analyze a broad mass range, making them suitable for complex mixtures and high-throughput analyses.

### 15.3.3 Orbitrap Mass Analyzer

The Orbitrap mass analyzer is a relatively recent innovation that combines high resolution and mass accuracy. In this design, ions are trapped in an electrostatic field and oscillate around a central axis. The frequency of oscillation is measured to determine the m/z ratio of the ions. Orbitraps are known for their exceptional mass resolution and sensitivity, making them ideal for proteomics, metabolomics, and other applications requiring detailed mass analysis.

## 15.4 APPLICATIONS OF MASS SPECTROMETRY

### 15.4.1 Proteomics

Mass spectrometry plays a crucial role in proteomics, the study of proteins and their functions. Techniques such as ESI-MS and MALDI-MS are commonly employed to analyze protein samples, identify post-translational modifications, and elucidate protein-protein interactions. The combination of mass spectrometry with liquid chromatography (LC-MS) allows for the analysis of complex protein mixtures, providing insights into cellular processes and disease mechanisms.

### 15.4.2 Metabolomics

Metabolomics is the study of small molecules, known as metabolites, within biological systems. Mass spectrometry is instrumental in identifying and quantifying metabolites in various biological samples, including urine, blood, and tissues. By analyzing metabolite profiles, researchers can gain insights into metabolic pathways, disease states, and the effects of drugs or environmental factors on metabolism.

### 15.4.3 Environmental Analysis

Mass spectrometry is widely used in environmental science to detect and quantify pollutants in air, water, and soil samples. Techniques such as GC-MS (gas chromatography-mass spectrometry) are employed to analyze volatile organic compounds, pesticides, and heavy metals. The sensitivity and specificity of mass spectrometry make it an invaluable tool for monitoring environmental contaminants and assessing their impact on human health and ecosystems.

### 15.4.4 Drug Discovery and Development

In the pharmaceutical industry, mass spectrometry is employed throughout the drug discovery and development process. It is used for characterizing drug candidates, studying pharmacokinetics, and analyzing metabolites. High-throughput mass spectrometry enables researchers to rapidly evaluate compound libraries and identify potential therapeutic agents. Additionally, mass spectrometry plays a critical role in ensuring the quality and safety of pharmaceutical products.

## 15.5 CHALLENGES AND FUTURE DIRECTIONS

### 15.5.1 Sensitivity and Detection Limits

One of the ongoing challenges in mass spectrometry is improving sensitivity and detection limits, especially for low-abundance analytes in complex biological matrices. Advances in ionization techniques, mass analyzers, and sample preparation methods are essential for enhancing sensitivity and enabling the detection of trace-level compounds.

### 15.5.2 Data Analysis and Interpretation

The analysis of mass spectrometry data can be complex, requiring sophisticated software and algorithms for accurate interpretation. Researchers must navigate challenges such as background noise, spectral overlap, and data processing. Continued development of data analysis tools will enhance the ability to extract meaningful information from mass spectrometry experiments.

### 15.5.3 Integration with Other Techniques

The future of mass spectrometry lies in its integration with other analytical techniques, such as chromatography, NMR, and imaging methods. Combining these approaches can provide comprehensive insights into molecular interactions, dynamics, and structures, leading to a deeper understanding of biological systems.

## 15.6 CONCLUSION

Mass spectrometric techniques have revolutionized the field of analytical chemistry, providing powerful tools for the analysis of complex mixtures and the elucidation of molecular structures. The versatility of mass spectrometry, combined with advancements in ionization techniques, mass analyzers, and data analysis, continues to expand its applications across diverse fields, including proteomics, metabolomics, environmental science, and drug development. As technology advances, mass spectrometry will remain at the forefront of scientific research, driving discoveries that enhance our understanding of the molecular world.

# CHAPTER 16
# FUNDAMENTALS OF BIOINFORMATICS

Bioinformatics is a multidisciplinary field that combines biology, computer science, mathematics, and statistics to analyze and interpret biological data. As the amount of biological information generated by high-throughput technologies continues to increase, the need for effective computational tools and techniques becomes paramount.

## 16.1 IMPORTANCE OF BIOINFORMATICS

The rapid advancement of genomic and proteomic technologies has led to the generation of vast amounts of biological data. This deluge of information presents both opportunities and challenges for researchers. Bioinformatics provides the means to store, analyze, and interpret this data, allowing for the extraction of meaningful biological insights. The importance of bioinformatics can be seen in several areas, including genomics, drug discovery, personalized medicine, and evolutionary biology.

### 16.1.1 Genomics and Data Analysis

In genomics, bioinformatics plays a crucial role in the analysis of DNA sequences. The Human Genome Project, which mapped the entire human genome, relied heavily on bioinformatics tools to assemble and annotate the sequence data. Researchers can use bioinformatics to identify genes, regulatory elements, and variants associated with diseases, paving the way for advancements in genetic research and diagnostics.

### 16.1.2 Drug Discovery and Development

Bioinformatics also significantly impacts drug discovery and development. By analyzing biological data, researchers can identify potential drug targets and predict the efficacy and safety of new compounds. Bioinformatics tools facilitate virtual screening of chemical libraries, helping to streamline the drug development process and reduce costs.

### 16.1.3 Personalized Medicine

The rise of personalized medicine—tailoring medical treatment to the individual characteristics of each patient—relies on bioinformatics for the integration of genomic information with clinical data. Bioinformatics enables the analysis of patient-specific genetic variations, allowing healthcare providers to make more informed treatment decisions based on an individual's genetic makeup.

## 16.2 CORE CONCEPTS OF BIOINFORMATICS

### 16.2.1 Sequence Analysis

In bioinformatics, sequence analysis is the process of subjecting a DNA, RNA or peptide sequence to any of a wide range of analytical methods to understand its features, function, structure, or evolution. It can be performed on the entire genome, transcriptome or proteome of an organism, and can also involve only selected segments or regions, like tandem repeats and transposable elements. Methodologies used include sequence alignment, searches against biological databases, and others.

Since the development of methods of high-throughput production of gene and protein sequences, the rate of addition of new sequences to the databases increased very rapidly. Such a collection of sequences does not, by itself, increase the scientist's understanding of the biology of organisms. However, comparing these new sequences to those with known functions is a key way of understanding the biology of an organism from which the new sequence comes.

Thus, sequence analysis can be used to assign function to coding and non-coding regions in a biological sequence usually by comparing sequences and studying similarities and differences. Nowadays, there are many tools and techniques that provide the sequence comparisons (sequence alignment) and analyze the alignment product to understand its biology.

Sequence analysis in molecular biology includes a very wide range of processes:

- The comparison of sequences to find similarity, often to infer if they are related (homologous).
- Identification of intrinsic features of the sequence such as active sites, post translational modification sites, gene-structures, reading frames, distributions of introns and exons and regulatory elements.
- Identification of sequence differences and variations such as point mutations and single nucleotide polymorphism (SNP) in order to get the genetic marker.
- Revealing the evolution and genetic diversity of sequences and organisms.
- Identification of molecular structure from sequence alone.

## History

Since the very first sequences of the insulin protein were characterized by Fred Sanger in 1951, biologists have been trying to use this knowledge to understand the function of molecules. He and his colleagues' discoveries contributed to the successful sequencing of the first DNA-based genome. The method used in this study, which is called the "Sanger method" or Sanger sequencing, was a milestone in sequencing long strand molecules such as DNA.

This method was eventually used in the human genome project. According to Michael Levitt, sequence analysis was born in the period from 1969 to 1977. In 1969 the analysis of sequences of transfer RNAs was used to infer residue interactions from correlated changes in the nucleotide sequences, giving rise to a model of the tRNA secondary structure. In 1970, Saul B. Needleman and Christian D. Wunsch published the first computer algorithm for aligning two sequences. Over this time, developments in obtaining nucleotide sequence improved greatly, leading to the publication of the first complete genome of a bacteriophage in 1977. Robert Holley and his team in Cornell University were believed to be the first to sequence an RNA molecule.

## Overview of Nucleotide Sequence Analysis (DNA & RNA)

Nucleotide sequence analyses identify functional elements like protein binding sites, uncover genetic variations like SNPs, study gene expression patterns, and understand the genetic basis of traits. It helps to understand mechanisms that contribute to processes like replication and transcription.

### Quality Control and Preprocessing

Quality control assesses the quality of sequencing reads obtained from the sequencing technology (e.g. Illumina). It is the first step in sequence analysis to limit wrong conclusions due to poor quality data. The tools used at this stage depend on the sequencing platform. For instance, FastQC checks the quality of short reads (including RNA sequences), Nanoplot or PycoQC are used for long read sequences (e.g. Nanopore sequence reads), and MultiQC aggregates the result of FastQC in a webpage format.

Quality control provides information such as read lengths, GC content, presence of adapter sequences (for short reads), and a quality score, which is often expressed on a PHRED scale. If adapters or other artifacts from PCR amplification are present in the reads (particularly short reads), they are removed using software such as Trimmomatic or Cutadapt.

### Read Alignment

At this step, sequencing reads whose quality have been improved are mapped to a reference genome using alignment tools like BWA for short DNA sequence reads, minimap for long read DNA sequences, and STAR for RNA sequence reads. The purpose of mapping is to find the origin of any given read based on the reference sequence. It is also important for detecting variations or phylogenetic studies. The output from this step, that is, the aligned reads, are stored in compatible file formats known as SAM, which contains information about the reference genome as well as individual reads. Alternatively, BAM file formats are preferred as they use much less desk or storage space.

### Variant Calling

Identifying variants is a popular aspect of sequence analysis as variants often contain information of biological significance, such as explaining the mechanism of drug resistance in an infectious disease. These variants could be single nucleotide variants (SNVs), small insertions/deletions (indels), and large structural variants. The read alignments are sorted using SAMtools, after which variant callers such as GATK are used to identify differences compared to the reference sequence.

The choice of variant calling tool depends heavily on the sequencing technology used, so GATK is often used when working with short reads, while long read sequences require tools like DeepVariant and Sniffles. Tools may also differ based on organism (prokaryotes or eukaryotes), source of sequence data (cancer vs metagenomic), and variant type of interest (SNVs or structural variants). The output of variant calling is typically in vcf format, and can be filtered using allele frequencies, quality scores, or other factors based on the research question at hand.

### Variant Annotation

This step adds context to the variant data using curated information from peer-reviewed papers and publicly available databases like gnomAD and Ensembl. Variants can be annotated with information about genomic features, functional consequences, regulatory elements, and population frequencies using tools like ANNOVAR or SnpEff, or custom scripts and pipeline. The output from this step is an annotation file in bed or txt format.

### Visualization and Interpretation

Genomic data, such as read alignments, coverage plots, and variant calls, can be visualized using genome browsers like IGV (Integrative Genomics Viewer) or UCSC Genome Browser. Interpretation of the results is done in the context of the biological question or hypothesis under investigation. The output can be a graphical representation of data in the forms of Circos plots, volcano plots, etc., or other forms of report describing the observations. DNA sequence analysis could also involve statistical modeling to infer relationships and epigenetic analysis, like identifying differential methylation regions using a tool like DSS.

### Gene Expression Analysis

Mapped RNA sequences are analyzed to estimate gene expression levels using quantification tools such as HTSeq, and identify differentially expressed genes (DEGs) between experimental conditions using statistical methods like DESeq2. This is carried out to compare the expression levels of genes or isoforms between or across different samples, and infer biological relevance. The output of gene expression analysis is typically a table with values representing the expression levels of gene IDs or names in rows and samples in the columns as well as standard errors and p-values. The results in the table can be further visualized using volcano plots and heatmaps, where colors represent the estimated expression level. Packages like ggplot2 in R and Matplotlib in Python are often used to create the visuals. The table can also be annotated using a reference annotation file, usually in GTF or GFF format to provide more context about the genes, such as the chromosome name, strand, and start and positions, and aid result interpretation.

### Functional Enrichment Analysis

Functional enrichment analysis identifies biological processes, pathways, and functional impacts associated with differentially expressed genes obtained from the previous step. It uses tools like GOSeq and Pathview. This creates a table with information about what pathways and molecular processes are associated with the differentially expressed genes, what genes are down or upregulated, and what gene ontology terms are recurrent or over-represented. RNA sequence analysis explores gene expression dynamics and regulatory mechanisms underlying biological processes and diseases. Interpretation of images and tables are carried out within the context of the hypotheses being investigated.

## Analyzing Protein Sequences

Proteome sequence analysis studies the complete set of proteins expressed by an organism or a cell under specific conditions. It describes protein structure, function, post-translational modifications, and interactions within biological systems. It often starts with raw mass spectrometry (MS) data from proteomics experiments, typically in mzML, mzXML, or RAW file formats.

Beyond preprocessing raw MS data to remove noise, normalize intensities, and detect peaks and converting proprietary file formats (e.g., RAW) to open-source formats (mzML, mzXML) for compatibility with downstream analysis tools, other analytical steps include peptide identification, peptide quantification, protein inference and quantification, generating quality control report, and normalization, imputation and significance testing. The choice and order of analytical steps depend on the MS method used, which can either be data dependent acquisition (DDA) or independent acquisition (DIA).

## Genome Browsers in Sequence Analysis

Genome browsers offer a non-code, user-friendly interface to visualize genomes and genomic segments, identify genomic features, and analyze the relationship between numerous genomic elements. The three primary genome browsers—Ensembl genome browser, UCSC genome browser, and the National Centre for Biotechnology Information (NCBI)—support different sequence analysis procedures, including genome assembly, genome annotation, and comparative genomics like exploring differential expression patterns and identifying conserved regions. All browsers support multiple data formats for upload and download and provide links to external tools and resources for sequence analyses, which contributes to their versatility.

## Sequence Alignment

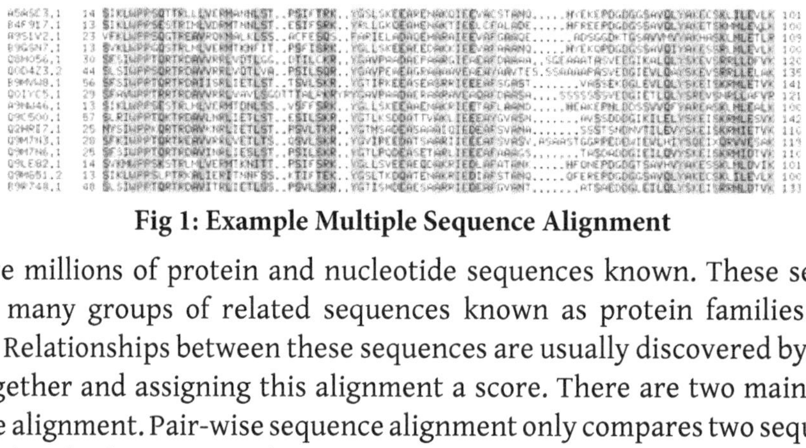

**Fig 1: Example Multiple Sequence Alignment**

There are millions of protein and nucleotide sequences known. These sequences fall into many groups of related sequences known as protein families or gene families. Relationships between these sequences are usually discovered by aligning them together and assigning this alignment a score. There are two main types of sequence alignment. Pair-wise sequence alignment only compares two sequences at a time and multiple sequence alignment compares many sequences. Two important algorithms for aligning pairs of sequences are the Needleman-Wunsch algorithm and the Smith-Waterman algorithm. Popular tools for sequence alignment include:

- **Pair-Wise Alignment:** BLAST, Dot plots.
- **Multiple Alignment:** ClustalW, PROBCONS, MUSCLE, MAFFT, and T-Coffee.

A common use for pairwise sequence alignment is to take a sequence of interest and compare it to all known sequences in a database to identify homologous sequences. In general, the matches in the database are ordered to show the most closely related sequences first, followed by sequences with diminishing similarity. These matches are usually reported with a measure of statistical significance such as an Expectation value.

## Profile Comparison

In 1987, Michael Gribskov, Andrew McLachlan, and David Eisenberg introduced the method of profile comparison for identifying distant similarities between proteins.

Rather than using a single sequence, profile methods use a multiple sequence alignment to encode a profile which contains information about the conservation level of each residue.

These profiles can then be used to search collections of sequences to find sequences that are related. Profiles are also known as Position Specific Scoring Matrices (PSSMs). In 1993, a probabilistic interpretation of profiles was introduced by Anders Krogh and colleagues using hidden Markov models. These models have become known as profile-HMMs. In recent years, methods have been developed that allow the comparison of profiles directly to each other. These are known as profile-profile comparison methods.

## Sequence Assembly

Sequence assembly refers to the reconstruction of a DNA sequence by aligning and merging small DNA fragments. It is an integral part of modern DNA sequencing. Since presently-available DNA sequencing technologies are ill-suited for reading long sequences, large pieces of DNA (such as genomes) are often sequenced by:

- Cutting the DNA into small pieces,
- Reading the small fragments, and
- Reconstituting the original DNA by merging the information on various fragments.

Recently, sequencing multiple species at one time is one of the top research objectives. Metagenomics is the study of microbial communities directly obtained from the environment. Different from cultured microorganisms from the lab, the wild sample usually contains dozens, sometimes even thousands of types of microorganisms from their original habitats. Recovering the original genomes can prove to be very challenging.

## Gene Prediction

Gene prediction or gene finding refers to the process of identifying the regions of genomic DNA that encode genes. This includes protein-coding genes as well as RNA genes, but may also include the prediction of other functional elements such as regulatory regions.

Geri is one of the first and most important steps in understanding the genome of a species once it has been sequenced. In general, the prediction of bacterial genes is significantly simpler and more accurate than the prediction of genes in eukaryotic species that usually have complex intron/exon patterns. Identifying genes in long sequences remains a problem, especially when the number of genes is unknown. Hidden markov models can be part of the solution. Machine learning has played a significant role in predicting the sequence of transcription factors. Traditional sequencing analysis focused on the statistical parameters of the nucleotide sequence itself. Another method is to identify homologous sequences based on other known gene sequences. The two methods described here are focused on the sequence. However, the shape feature of these molecules such as DNA and protein have also been studied and proposed to have an equivalent, if not higher, influence on the behaviors of these molecules.

## Protein Structure Prediction

The 3D structures of molecules are of major importance to their functions in nature. Since structural prediction of large molecules at an atomic level is a largely intractable problem, some biologists introduced ways to predict 3D structure at a primary sequence level. This includes the biochemical or statistical analysis of amino acid residues in local regions and structural the inference from homologs (or other potentially related proteins) with known 3D structures.

There have been a large number of diverse approaches to solve the structure prediction problem. In order to determine which methods were most effective, a structure prediction competition was founded called CASP (Critical Assessment of Structure Prediction).

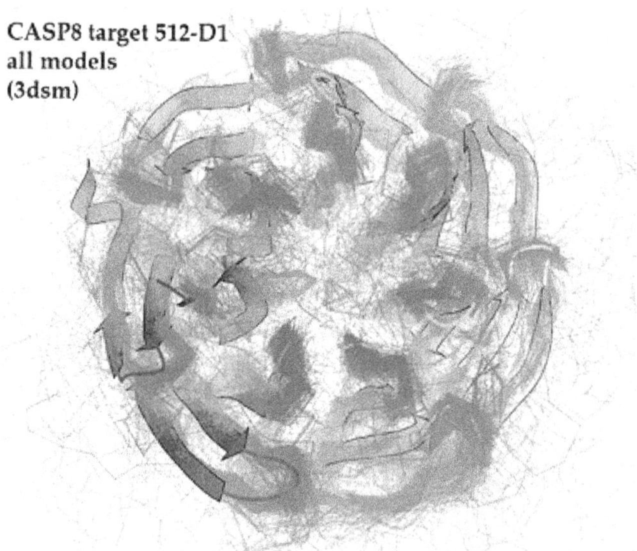


**Fig 2: Target Protein Structure (3dsm, shown in Ribbons), with Calpha Backbones (in Gray) of 354 Predicted Models for It Submitted in the CASP8 Structure-Prediction Experiment**

## Computational Approaches and Techniques

Sequence analysis tasks are often non-trivial to resolve and require the use of relatively complex approaches, many of which are the backbone behind many existing sequence analysis tools. Of the many methods used in practice, the most popular include the following:

- Dynamic programming.
- Artificial neural network.
- Hidden Markov model.
- Support vector machine.
- Clustering.
- Bayesian network.
- Regression analysis.
- Sequence mining.
- Alignment-free sequence analysis.

### 16.2.2 Structural Bioinformatics

Structural bioinformatics is the branch of bioinformatics that is related to the analysis and prediction of the three-dimensional structure of biological macromolecules such as proteins, RNA, and DNA. It deals with generalizations about macromolecular 3D structures such as comparisons of overall folds and local motifs, principles of molecular folding, evolution, binding interactions, and structure/function relationships, working both from experimentally solved structures and from computational models. The term *structural* has the same meaning as in structural biology, and structural bioinformatics can be seen as a part of computational structural biology. The main objective of structural bioinformatics is the creation of new methods of analysing and manipulating biological macromolecular data in order to solve problems in biology and generate new knowledge.

#### Introduction

#### Protein Structure

The structure of a protein is directly related to its function. The presence of certain chemical groups in specific locations allows proteins to act as enzymes, catalyzing several chemical reactions. In general, protein structures are classified into four levels: primary (sequences), secondary (local conformation of the polypeptide chain), tertiary (three-dimensional structure of the protein fold), and quaternary (association of multiple polypeptide structures).

Structural bioinformatics mainly addresses interactions among structures taking into consideration their space coordinates. Thus, the primary structure is better analyzed in traditional branches of bioinformatics. However, the sequence implies restrictions that allow the formation of conserved local conformations of the polypeptide chain, such as alpha-helix, beta-sheets, and loops (secondary structure). Also, weak interactions (such as hydrogen bonds) stabilize the protein fold. Interactions could be intrachain, i.e., when occurring between parts of the same protein monomer (tertiary structure), or interchain, i.e., when occurring between different structures (quaternary structure). Finally, the topological arrangement of interactions, whether strong or weak, and entanglements is being studied in the field of structural bioinformatics, utilizing frameworks such as circuit topology.

### Structure Visualization

Protein structure visualization is an important issue for structural bioinformatics. It allows users to observe static or dynamic representations of the molecules, also allowing the detection of interactions that may be used to make inferences about molecular mechanisms. The most common types of visualization are:

- **Cartoon:** This type of protein visualization highlights the secondary structure differences. In general, α-helix is represented as a type of screw, β-strands as arrows, and loops as lines.
- **Lines:** Each amino acid residue is represented by thin lines, which allows a low cost for graphic rendering.
- **Surface:** In this visualization, the external shape of the molecule is shown.
- **Sticks:** Each covalent bond between amino acid atoms is represented as a stick. This type of visualization is most used to visualize interactions between amino acids...

### DNA Structure

The classic DNA duplexes structure was initially described by Watson and Crick (and contributions of Rosalind Franklin). The DNA molecule is composed of three substances: a phosphate group, a pentose, and a nitrogen base (adenine, thymine, cytosine, or guanine). The DNA double helix structure is stabilized by hydrogen bonds formed between base pairs: adenine with thymine (A-T) and cytosine with guanine (C-G). Many structural bioinformatics studies have focused on understanding interactions between DNA and small molecules, which has been the target of several drug design studies.

### Interactions

Interactions are contacts established between parts of molecules at different levels. They are responsible for stabilizing protein structures and perform a varied range of activities. In biochemistry, interactions are characterized by the proximity of atom groups or molecules regions that present an effect upon one another, such as electrostatic forces, hydrogen bonding, and hydrophobic effect. Proteins can perform several types of interactions, such as protein-protein interactions (PPI), protein-peptide interactions, protein-ligand interactions (PLI), and protein-DNA interaction.

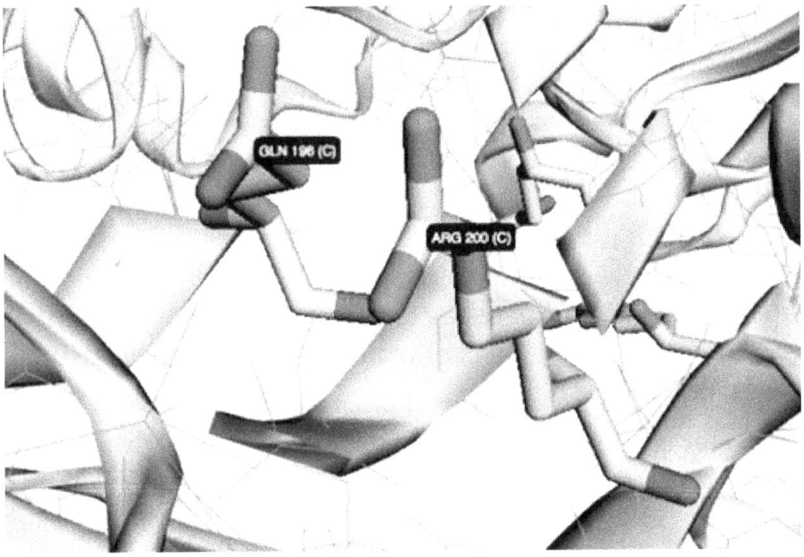


**Fig 3: Contacts between Two Amino Acid Residues: Q196-R200 (PDB ID- 2X1C)**

## Calculating Contacts

Calculating contacts is an important task in structural bioinformatics, being important for the correct prediction of protein structure and folding, thermodynamic stability, protein-protein and protein-ligand interactions, docking and molecular dynamics analyses, and so on. Traditionally, computational methods have used threshold distance between atoms (also called cutoff) to detect possible interactions. This detection is performed based on Euclidean distance and angles between atoms of determined types. However, most of the methods based on simple Euclidean distance cannot detect occluded contacts. Hence, cutoff free methods, such as Delaunay triangulation, have gained prominence in recent years. In addition, the combination of a set of criteria, for example, physicochemical properties, distance, geometry, and angles, have been used to improve the contact determination.

**Table 1: Distance Criteria for Contact Definition**

| Type | Max Distance Criteria |
|---|---|
| Hydrogen bond | 3,9 Å |
| Hydrophobic interaction | 5 Å |
| Ionic interaction | 6 Å |
| Aromatic Stacking | 6 Å |

## Protein Data Bank (PDB)

The Protein Data Bank (PDB) is a database of 3D structure data for large biological molecules, such as proteins, DNA, and RNA. PDB is managed by an international organization called the Worldwide Protein Data Bank (wwPDB), which is composed of several local organizations, as. PDBe, PDBj, RCSB, and BMRB. They are responsible for keeping copies of PDB data available on the internet at no charge. The number of structure data available at PDB has increased each year, being obtained typically by X-ray crystallography, NMR spectroscopy, or cryo-electron microscopy.

### Data Format

The PDB format (.pdb) is the legacy textual file format used to store information of three-dimensional structures of macromolecules used by the Protein Data Bank. Due to restrictions in the format structure conception, the PDB format does not allow large structures containing more than 62 chains or 99999 atom records.

The PDBx/mmCIF (macromolecular Crystallographic Information File) is a standard text file format for representing crystallographic information. Since 2014, the PDB format was substituted as the standard PDB archive distribution by the PDBx/mmCIF file format (.cif). While PDB format contains a set of records identified by a keyword of up to six characters, the PDBx/mmCIF format uses a structure based on key and value, where the key is a name that identifies some feature and the value is the variable information.

### Other Structural Databases

In addition to the Protein Data Bank (PDB), there are several databases of protein structures and other macromolecules. Examples include:

- **MMDB:** Experimentally determined three-dimensional structures of biomolecules derived from Protein Data Bank (PDB).
- **Nucleic Acid Data Base (NDB):** Experimentally determined information about nucleic acids (DNA, RNA).
- **Structural Classification of Proteins (SCOP):** Comprehensive description of the structural and evolutionary relationships between structurally known proteins.
- **TOPOFIT-DB:** Protein structural alignments based on the TOPOFIT method.
- **Electron Density Server (EDS):** Electron-density maps and statistics about the fit of crystal structures and their maps.
- **CASP:** Prediction Center Community-wide, worldwide experiment for protein structure prediction CASP.

- **PISCES Server for Creating Non-Redundant Lists of Proteins:** Generates PDB list by sequence identity and structural quality criteria.
- **The Structural Biology Knowledgebase:** Tools to aid in protein research design.
- **ProtCID:** The Protein Common Interface Database Database of similar protein-protein interfaces in crystal structures of homologous proteins.
- **AlphaFold:** AlphaFold - Protein Structure Database.

## Structure Comparison

### Structural Alignment

Structural alignment is a method for comparison between 3D structures based on their shape and conformation. It could be used to infer the evolutionary relationship among a set of proteins even with low sequence similarity. Structural alignment implies superimposing a 3D structure over a second one, rotating and translating atoms in corresponding positions (in general, using the $C_\alpha$ atoms or even the backbone heavy atoms $C$, $N$, $O$, and $C_\alpha$). Usually, the alignment quality is evaluated based on the root-mean-square deviation (RMSD) of atomic positions, *i.e.*, the average distance between atoms after superimposition:

$$\text{RMSD} = \sqrt{\frac{1}{N}\sum_{i=1}^{N}\delta_i^2}$$

where $\delta_i$ is the distance between atom *i* and either a reference atom corresponding in the other structure or the mean coordinate of the *N* equivalent atoms. In general, the RMSD outcome is measured in Ångström (Å) unit, which is equivalent to $^{10-}10$ m. The nearer to zero the RMSD value, the more similar are the structures.

### Graph-Based Structural Signatures

Structural signatures, also called fingerprints, are macromolecule pattern representations that can be used to infer similarities and differences. Comparisons among a large set of proteins using RMSD still is a challenge due to the high computational cost of structural alignments. Structural signatures based on graph distance patterns among atom pairs have been used to determine protein identifying vectors and to detect non-trivial information. Furthermore, linear algebra and machine learning can be used for clustering protein signatures, detecting protein-ligand interactions, predicting ΔΔG, and proposing mutations based on Euclidean distance.

## Structure Prediction

The atomic structures of molecules can be obtained by several methods, such as

X-ray crystallography (XRC), NMR spectroscopy, and 3D electron microscopy; however, these processes can present high costs and sometimes some structures can be hardly established, such as membrane proteins. Hence, it is necessary to use computational approaches for determining 3D structures of macromolecules. The structure prediction methods are classified into comparative modeling and de novo modeling.

### Comparative Modeling

Comparative modeling, also known as homology modeling, corresponds to the methodology to construct three-dimensional structures from an amino acid sequence of a target protein and a template with known structure. The literature has described that evolutionarily related proteins tend to present a conserved three-dimensional structure. In addition, sequences of distantly related proteins with identity lower than 20% can present different folds.

### De Novo Modeling

In structural bioinformatics, *de novo* modeling, also known as *ab initio* modeling, refers to approaches for obtaining three-dimensional structures from sequences without the necessity of a homologous known 3D structure. Despite the new algorithms and methods proposed in the last years, de novo protein structure prediction is still considered one of the remain outstanding issues in modern science.

### Structure Validation

After structure modeling, an additional step of structure validation is necessary since many of both comparative and 'de novo' modeling algorithms and tools use heuristics to try assembly the 3D structure, which can generate many errors. Some validation strategies consist of calculating energy scores and comparing them with experimentally determined structures. For example, the DOPE score is an energy score used by the MODELLER tool for determining the best model.

Another validation strategy is calculating $\phi$ and $\psi$ backbone dihedral angles of all residues and construct a Ramachandran plot. The side-chain of amino acids and the nature of interactions in the backbone restrict these two angles, and thus, the visualization of allowed conformations could be performed based on the Ramachandran plot. A high quantity of amino acids allocated in no permissive positions of the chart is an indication of a low-quality modeling.

### Prediction Tools

A list with commonly used software tools for protein structure prediction, including comparative modeling, protein threading, *de novo* protein structure prediction, and secondary structure prediction is available in the list of protein structure prediction software.

## Molecular Docking

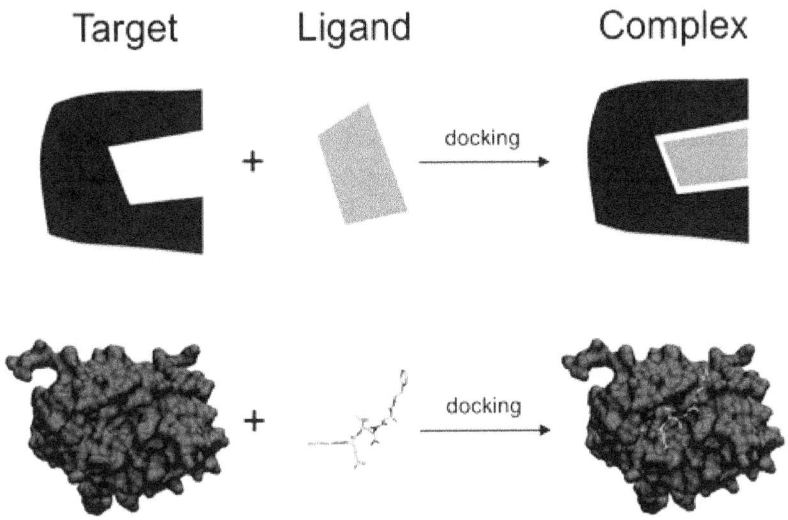


**Fig 4: Representation of Docking a Ligand (Green) to a Protein Target (Black)**

Molecular docking is a method used to predict the orientation coordinates of a molecule (ligand) when bound to another one (receptor or target). The binding may be mostly through non-covalent interactions while covalently linked binding can also be studied.

Molecular docking aims to predict possible poses (binding modes) of the ligand when it interacts with specific regions on the receptor. Docking tools use force fields to estimate a score for ranking best poses that favored better interactions between the two molecules.

In general, docking protocols are used to predict the interactions between small molecules and proteins. However, docking also can be used to detect associations and binding modes among proteins, peptides, DNA or RNA molecules, carbohydrates, and other macromolecules.

### Virtual Screening

Virtual screening (VS) is a computational approach used for fast screening of large compound libraries for drug discovery. Usually, virtual screening uses docking algorithms to rank small molecules with the highest affinity to a target receptor.

In recent times, several tools have been used to evaluate the use of virtual screening in the process of discovering new drugs. However, problems such as missing information, inaccurate understanding of drug-like molecular properties, weak scoring functions, or insufficient docking strategies hinder the docking process. Hence, the literature has described that it is still not considered a mature technology.

## Molecular Dynamics

Molecular dynamics (MD) is a computational method for simulating interactions between molecules and their atoms during a given period of time.

This method allows the observation of the behavior of molecules and their interactions, considering the system as a whole. To calculate the behavior of the systems and, thus, determine the trajectories, an MD can use Newton's equation of motion, in addition to using molecular mechanics methods to estimate the forces that occur between particles (force fields).

## Applications

Informatics approaches used in structural bioinformatics are:

- **Selection of Target:** Potential targets are identified by comparing them with databases of known structures and sequence. The importance of a target can be decided on the basis of published literature. Target can also be selected on the basis of its protein domain. Protein domains are building blocks that can be rearranged to form new proteins. They can be studied in isolation initially.
- **Tracking X-Ray Crystallography Trials:** X-Ray crystallography can be used to reveal three-dimensional structure of a protein. But, in order to use X-ray for studying protein crystals, pure proteins crystals must be formed, which can take a lot of trials. This leads to a need for tracking the conditions and results of trials. Furthermore, supervised machine learning algorithms can be used on the stored data to identify conditions that might increase the yield of pure crystals.
- **Analysis of X-Ray Crystallographic Data:** The diffraction pattern obtained as a result of bombarding X-rays on electrons is Fourier transform of electron density distribution. There is a need for algorithms that can deconvolve Fourier transform with partial information ( due to missing phase information, as the detectors can only measure amplitude of diffracted X-rays, and not the phase shifts ). Extrapolation technique such as Multiwavelength anomalous dispersion can be used to generate electron density map, which uses the location of selenium atoms as a reference to determine rest of the structure. Standard Ball-and-stick model is generated from the electron density map.
- **Analysis of NMR Spectroscopy Data:** Nuclear magnetic resonance spectroscopy experiments produce two (or higher) dimensional data, with each peak corresponding to a chemical group within the sample. Optimization methods are used to convert spectra into three dimensional structures.

- **Correlating Structural Information with Functional Information:** Structural studies can be used as probe for structural-functional relationship.

## Tools

**Table 2: List of Structural Bioinformatics Tools**

| Software | Description |
|---|---|
| I-TASSER | Predicting three-dimensional structure model of protein molecules from amino acid sequences. |
| MOE | Molecular Operating Environment (MOE) is an extensive platform including structural modeling for proteins, protein families and antibodies |
| SBL | The Structural Bioinformatics Library: end-user applications and advanced algorithms |
| BALLView | Molecular modeling and visualization |
| STING | Visualization and analysis |
| PyMOL | Viewer and modeling |
| VMD | Viewer, molecular dynamics |
| Gromacs | Protein folding, molecular dynamics, molecular model refinement, molecular model force field generation |
| LAMMPS | Protein folding, molecular dynamics, molecular model refinement, Quantum mechanical macro-molecular interactions |
| GAMESS | Molecular Force Field, Charge refinement, Quantum molecular dynamics, Protein-Molecular chemical reaction simulations (electron transfer), |
| KiNG | An open-source Java kinemage viewer |
| STRIDE | Determination of secondary structure from coordinates |
| DSSP | Algorithm assigning a secondary structure to the amino acids of a protein |
| MolProbity | Structure-validation web server |
| PROCHECK | A structure-validation web service |
| CheShift | A protein structure-validation on-line application |
| 3D-mol.js | A molecular viewer for web applications developed using Javascript |
| PROPKA | Rapid prediction of protein pKa values based on empirical structure/ function relationships |
| CARA | Computer Aided Resonance Assignment |
| Docking Server | A molecular docking web server |
| StarBiochem | A java protein viewer, features direct search of protein databank |
| SPADE | The structural proteomics application development environment |

| Software | Description |
|---|---|
| PocketSuite | A web portal for various web-servers for binding site-level analysis. PocketSuite is divided into:: PocketDepth (Binding site prediction)<br>PocketMatch (Binding site comparison), PocketAlign (Binding site alignment), and PocketAnnotate (Binding site annotation). |
| MSL | An open-source C++ molecular modeling software library for the implementation of structural analysis, prediction and design methods |
| PSSpred | Protein secondary structure prediction |
| Proteus | Webtool for suggesting mutation pairs |
| SDM | A server for predicting effects of mutations on protein stability |

### 16.2.3 Genomics and Transcriptomics

Genomics refers to the study of an organism's entire genome, while transcriptomics focuses on the analysis of gene expression profiles. Bioinformatics tools are used to analyze high-throughput sequencing data from technologies such as RNA-Seq, which quantifies RNA levels in different conditions. By examining gene expression patterns, researchers can gain insights into cellular responses to various stimuli and the regulatory mechanisms governing gene expression.

### 16.2.4 Proteomics

Proteomics involves the large-scale study of proteins, including their structure, function, and interactions. Bioinformatics plays a key role in proteomics by enabling the analysis of mass spectrometry data and protein-protein interaction networks. Tools like MaxQuant and STRING help researchers identify proteins and their interactions, facilitating a deeper understanding of cellular processes and pathways.

## 16.3 TOOLS AND TECHNIQUES IN BIOINFORMATICS

### 16.3.1 Databases

Bioinformatics relies heavily on databases that store biological information. These databases, such as GenBank, UniProt, and the Protein Data Bank (PDB), provide researchers with access to a wealth of sequence and structural data. Researchers can query these databases to retrieve relevant information for their studies, making data sharing and collaboration more efficient.

### 16.3.2 Software and Algorithms

Various software tools and algorithms are developed to perform specific bioinformatics tasks. For sequence alignment, tools like MUSCLE and Bowtie are widely used, while tools such as GROMACS and NAMD are employed for molecular dynamics simulations. Machine learning and artificial intelligence techniques are increasingly being integrated into bioinformatics, enabling the development of predictive models for biological phenomena.

### 16.3.3 Data Visualization

Data visualization is essential in bioinformatics, as it allows researchers to interpret complex data sets effectively.

Tools such as R and Python libraries (e.g., Matplotlib, Seaborn) are frequently used to create visual representations of biological data, such as heatmaps, scatter plots, and network diagrams.

Visualization aids in the identification of patterns and trends, enhancing the understanding of biological systems.

## 16.4 APPLICATIONS OF BIOINFORMATICS

### 16.4.1 Comparative Genomics

Comparative genomics is the analysis of genome evolution by comparing the genomes of different species.

Bioinformatics tools facilitate the identification of conserved genes, genomic rearrangements, and evolutionary relationships. By studying these aspects, researchers can gain insights into the evolutionary history of organisms and the functional significance of specific genes.

### 16.4.2 Systems Biology

Systems biology is an interdisciplinary field that focuses on understanding complex biological systems through computational modeling and analysis. Bioinformatics plays a crucial role in integrating data from various biological sources, including genomics, transcriptomics, proteomics, and metabolomics. By constructing computational models of biological pathways and networks, researchers can simulate and predict cellular behavior under different conditions.

### 16.4.3 Metagenomics

Metagenomics involves the study of genetic material recovered directly from environmental samples, providing insights into the diversity and functionality of microbial communities. Bioinformatics tools enable the analysis of metagenomic data, allowing researchers to identify microbial species, predict metabolic functions, and understand the interactions within microbial ecosystems.

### 16.4.4 Vaccine Development

Bioinformatics has become increasingly important in vaccine development, particularly in the identification of potential vaccine candidates. By analyzing genomic data from pathogens, researchers can identify conserved antigens and evaluate their immunogenic potential. Bioinformatics tools also aid in the design of novel vaccines through in silico modeling and simulation of immune responses.

## 16.5 FUTURE PROSPECTS OF BIOINFORMATICS

### 16.5.1 Integration of Omics Technologies

The future of bioinformatics lies in the integration of various omics technologies, including genomics, transcriptomics, proteomics, and metabolomics. By combining data from multiple sources, researchers can obtain a comprehensive understanding of biological systems, paving the way for new discoveries in fields such as personalized medicine and systems biology.

### 16.5.2 Advancements in Artificial Intelligence

The integration of artificial intelligence (AI) and machine learning into bioinformatics is expected to revolutionize the field. AI algorithms can analyze vast datasets, identify patterns, and make predictions that would be challenging for humans to discern. This technology holds great promise for improving drug discovery, disease diagnosis, and the understanding of complex biological processes.

### 16.5.3 Ethical Considerations

As bioinformatics continues to advance, ethical considerations surrounding data privacy and the use of biological information will become increasingly important. The responsible use of genomic and health data is crucial to ensure that individuals' privacy is protected while still allowing for valuable research and innovation.

## 16.6 CONCLUSION

Bioinformatics is a dynamic and rapidly evolving field that plays a critical role in modern biological research. By integrating biological data with computational tools and techniques, bioinformatics enables researchers to analyze, interpret, and draw meaningful conclusions from complex datasets. Its applications span various domains, from genomics and proteomics to drug discovery and systems biology. As technology continues to advance, bioinformatics will remain at the forefront of scientific inquiry, providing essential insights into the molecular mechanisms of life and driving innovations in medicine and biotechnology.

# CHAPTER 17

# FUNDAMENTALS OF CHEMOINFORMATICS

Chemoinformatics, also known as cheminformatics, is a rapidly evolving field that intersects chemistry, computer science, and information technology. It encompasses the application of computational techniques and tools to solve problems in chemical research, facilitate drug discovery, and enhance our understanding of molecular properties and behavior. The importance of cheminformatics is underscored by the ever-increasing volume of chemical data generated in the modern era of high-throughput experimentation.

## 17.1 DEFINITION AND SCOPE OF CHEMOINFORMATICS

Chemoinformatics can be defined as the use of computational methods to analyze, visualize, and predict chemical phenomena. It integrates various disciplines, including organic chemistry, molecular biology, computer science, and statistics, to manage and interpret chemical data. The scope of cheminformatics extends beyond mere data analysis; it encompasses the entire pipeline from data collection and storage to analysis, modeling, and prediction.

In the context of cheminformatics, data can include chemical structures, biological activities, physicochemical properties, and spectral data. By utilizing advanced computational tools and algorithms, researchers can process this data to gain insights into chemical behavior, optimize chemical compounds, and streamline the drug discovery process.

## 17.2 CORE CONCEPTS OF CHEMOINFORMATICS

### 17.2.1 Chemical Representation

A fundamental concept in cheminformatics is the representation of chemical structures. Chemicals can be represented in various ways, including molecular formulas, structural diagrams, and three-dimensional (3D) coordinates. The most common representation is the use of line-angle structural formulas, where atoms are represented as symbols (e.g., C for carbon, O for oxygen) and bonds are depicted as lines.

Additionally, cheminformatics employs various file formats for storing chemical information, such as SMILES (Simplified Molecular Input Line Entry System) and

InChI (International Chemical Identifier). These formats provide a way to encode molecular structures in a textual format, making it easier to share and analyze chemical data across different software platforms.

### 17.2.2 Data Mining and Analysis

Data mining is a crucial aspect of cheminformatics, involving the extraction of meaningful patterns and insights from large chemical datasets. Researchers use statistical methods and machine learning algorithms to identify correlations and trends in chemical properties, activities, and behaviors. This analysis allows for the development of predictive models that can forecast the behavior of new compounds based on existing data.

### 17.2.3 Molecular Descriptors

Molecular descriptors are quantitative representations of molecular properties that can be used to characterize chemical compounds. They can be classified into different categories, including constitutional descriptors (based on the molecular structure), topological descriptors (based on the connectivity of atoms), and geometric descriptors (based on molecular shape).

These descriptors play a critical role in cheminformatics as they serve as input features for various predictive modeling approaches. By correlating molecular descriptors with biological activities or physicochemical properties, researchers can develop quantitative structure-activity relationship (QSAR) models that predict the behavior of untested compounds.

## 17.3 TECHNIQUES IN CHEMOINFORMATICS

### 17.3.1 Structure-Activity Relationship (SAR) Analysis

Structure-activity relationship (SAR) analysis is a foundational technique in cheminformatics that examines the relationship between the chemical structure of compounds and their biological activity. By analyzing known compounds with defined activities, researchers can identify key structural features that contribute to their effectiveness. This information guides the design of new compounds with improved efficacy and reduced toxicity.

### 17.3.2 Molecular Docking

Molecular docking is a computational technique used to predict the preferred orientation of a small molecule (ligand) when bound to a target protein (receptor). This method is crucial in drug discovery as it helps identify potential drug candidates and understand their interactions with biological targets. Molecular docking simulations provide insights into binding affinities, binding modes, and the effects of structural modifications on activity.

### 17.3.3 Virtual Screening

Virtual screening is an essential tool in cheminformatics that involves the computational screening of large compound libraries to identify potential hits with desired biological activity. This process typically employs scoring functions to evaluate the binding affinity of compounds to a target, allowing researchers to prioritize candidates for further experimental testing. Virtual screening accelerates the drug discovery process by reducing the number of compounds that need to be synthesized and tested.

### 17.3.4 Chemoinformatics Databases

Chemoinformatics relies heavily on databases that store chemical information and facilitate data retrieval and analysis. Numerous databases, such as PubChem, ChEMBL, and the Cambridge Structural Database (CSD), provide access to a wealth of chemical data, including molecular structures, biological activities, and physicochemical properties. Researchers can query these databases to find relevant information for their studies, enabling more informed decision-making during the research process.

## 17.4 APPLICATIONS OF CHEMOINFORMATICS

### 17.4.1 Drug Discovery and Development

One of the most significant applications of cheminformatics is in the field of drug discovery and development. Chemoinformatics tools and techniques streamline the identification of potential drug candidates by enabling researchers to analyze large chemical libraries and predict the activity of new compounds. This approach reduces the time and costs associated with traditional drug development methods.

### 17.4.2 Toxicology Assessment

Chemoinformatics is also instrumental in predicting the toxicological properties of compounds. By analyzing existing data on chemical toxicity, researchers can develop predictive models that assess the potential hazards of new substances. This information is critical for regulatory compliance and safety assessments, ensuring that new chemicals are evaluated for their potential risks before reaching the market.

### 17.4.3 Materials Science

In addition to drug discovery, cheminformatics finds applications in materials science. Researchers utilize cheminformatics tools to design and optimize new materials with specific properties, such as polymers, catalysts, and nanomaterials. By understanding the relationship between molecular structure and material performance, researchers can develop advanced materials for various applications, including energy storage and environmental remediation.

### 17.4.4 Environmental Chemistry

Chemoinformatics plays a vital role in environmental chemistry by aiding in the assessment of chemical pollutants and their impact on ecosystems. Researchers can model the fate and transport of contaminants in the environment, evaluate their toxicity to organisms, and develop remediation strategies. This information is essential for environmental protection and regulatory compliance.

## 17.5 FUTURE PROSPECTS OF CHEMOINFORMATICS

### 17.5.1 Integration of Artificial Intelligence

The future of cheminformatics is closely tied to the integration of artificial intelligence (AI) and machine learning techniques. These technologies have the potential to revolutionize the analysis of chemical data by automating data mining, enhancing predictive modeling, and improving decision-making processes.

AI algorithms can analyze complex datasets at unprecedented speeds, uncovering hidden patterns that may be challenging for humans to discern.

### 17.5.2 Personalized Medicine

As personalized medicine continues to gain traction, cheminformatics will play a crucial role in tailoring drug treatments to individual patients. By analyzing genetic, proteomic, and metabolomic data alongside chemical information, researchers can develop targeted therapies that are more effective and have fewer side effects. This integration of cheminformatics with clinical data will pave the way for more precise and individualized medical treatments.

### 17.5.3 Open Science and Data Sharing

The trend towards open science and data sharing is expected to enhance collaboration and innovation in cheminformatics. By making chemical data freely accessible to researchers worldwide, the field can benefit from collective knowledge and resources. Initiatives promoting open data and collaborative platforms will facilitate the exchange of information, leading to faster advancements in chemical research and applications.

### 17.5.4 Ethical Considerations

As cheminformatics continues to evolve, ethical considerations surrounding data privacy, intellectual property, and the responsible use of chemical information will become increasingly important. Researchers must navigate the ethical implications of data sharing and ensure that research practices align with ethical standards.

Responsible data management and transparency will be essential for maintaining public trust in the field.

## 17.6 CONCLUSION

Chemoinformatics is a vital and dynamic discipline that bridges the gap between chemistry and computational science. By leveraging computational tools and techniques, researchers can analyze and interpret complex chemical data, leading to significant advancements in drug discovery, materials science, and environmental chemistry.

As the field continues to evolve, the integration of artificial intelligence, open science initiatives, and ethical considerations will shape the future of cheminformatics, fostering innovation and collaboration in the scientific community. The ongoing development of cheminformatics promises to enhance our understanding of chemical systems and facilitate the design of new compounds, ultimately improving human health and the environment.

# CHAPTER 18

# FUNDAMENTALS OF GENOME SEQUENCING AND ANNOTATION

Genome sequencing and annotation are pivotal processes in the field of genomics, providing insights into the genetic blueprint of organisms. With the advancement of technology, these processes have become more efficient and accessible, transforming our understanding of biology, evolution, and disease.

## 18.1 UNDERSTANDING GENOME SEQUENCING

### 18.1.1 Definition and Purpose

Genome sequencing is the process of determining the complete nucleotide sequence of an organism's DNA. It involves identifying the order of the four nucleotide bases—adenine (A), thymine (T), cytosine (C), and guanine (G)—in the genome. The primary purpose of genome sequencing is to provide a comprehensive map of an organism's genetic material, which serves as a foundation for various biological and medical applications.

### 18.1.2 Historical Background

The journey of genome sequencing began in the late 20th century, culminating in the Human Genome Project (HGP), which was initiated in 1990 and completed in 2003. The HGP aimed to sequence the entire human genome, comprising approximately 3 billion base pairs. This monumental project set the stage for modern genomics, paving the way for subsequent sequencing initiatives across diverse species, including plants, animals, and microorganisms.

## 18.2 TYPES OF GENOME SEQUENCING TECHNIQUES

Genome sequencing techniques have evolved significantly, leading to the development of various methodologies. These can be broadly categorized into two main types: Sanger sequencing and next-generation sequencing (NGS).

### 18.2.1 Sanger Sequencing

Sanger sequencing, also known as chain termination sequencing, was developed by Frederick Sanger in the 1970s. It involves the use of labeled dideoxynucleotides to terminate DNA strand elongation during replication, generating fragments of varying lengths. The fragments are then separated by size using capillary electrophoresis, allowing the determination of the nucleotide sequence. Although Sanger sequencing is highly accurate, it is relatively slow and costly for large-scale projects, making it less suitable for whole-genome sequencing.

### 18.2.2 Next-Generation Sequencing (NGS)

Next-generation sequencing (NGS) has revolutionized genome sequencing by enabling high-throughput and rapid sequencing of entire genomes. NGS technologies, such as Illumina, Ion Torrent, and Pacific Biosciences, allow millions of DNA fragments to be sequenced simultaneously, significantly reducing both time and cost. NGS involves several steps, including library preparation, sequencing, and data analysis. The advent of NGS has facilitated large-scale genomic studies, including comparative genomics, metagenomics, and personalized medicine.

### 18.2.3 Third-Generation Sequencing

Third-generation sequencing technologies, such as single-molecule real-time (SMRT) sequencing and nanopore sequencing, offer the ability to read longer DNA fragments and provide real-time sequencing data. These advancements enhance the accuracy of genomic assembly and enable the identification of structural variations and epigenetic modifications. Third-generation sequencing holds promise for more comprehensive genomic analyses and a deeper understanding of complex genomes.

## 18.3 GENOME ANNOTATION

### 18.3.1 Definition and Importance

Genome annotation is the process of identifying and labeling the functional elements within a genome, such as genes, regulatory regions, and non-coding sequences. Annotation provides context to the raw sequence data generated through sequencing, allowing researchers to understand the roles and relationships of different genomic components. Accurate genome annotation is crucial for deciphering biological functions, studying evolutionary relationships, and facilitating biomedical research.

### 18.3.2 Types of Genome Annotation

Genome annotation can be classified into two main types: structural annotation and functional annotation.

#### Structural Annotation

Structural annotation involves identifying the physical features of the genome, including gene locations, exon-intron structures, and repetitive elements. This process typically employs computational tools and algorithms that utilize known gene sequences and homology searches against reference genomes. Structural annotation serves as the foundation for subsequent functional analysis, providing a framework for understanding gene organization.

### Functional Annotation

Functional annotation assigns biological functions to the identified genes and genomic elements.

This process involves comparing the sequenced genome against databases of known genes, proteins, and pathways to predict functions based on sequence similarity.

Functional annotation may also incorporate experimental data, such as gene expression profiles and protein interactions, to enhance the understanding of gene functions within specific biological contexts.

## 18.4 METHODS AND TOOLS FOR GENOME ANNOTATION

Various methods and tools are employed in genome annotation, ranging from computational algorithms to experimental approaches.

### 18.4.1 Computational Tools

Computational tools play a crucial role in genome annotation, leveraging algorithms to analyze sequence data and predict gene functions. Commonly used tools include:

- **BLAST (Basic Local Alignment Search Tool):** BLAST is a widely used algorithm for comparing nucleotide or protein sequences against databases to identify homologous sequences. It facilitates the identification of conserved genes and functional domains.
- **AUGUSTUS:** AUGUSTUS is a gene prediction software that employs hidden Markov models to predict gene structures based on sequence data. It incorporates known gene models and allows for species-specific customization.
- **GeneMark:** GeneMark is another gene prediction tool that uses statistical models to identify coding regions within genomic sequences. It is particularly effective for prokaryotic genomes and has been adapted for eukaryotic annotations.

### 18.4.2 Experimental Approaches

In addition to computational methods, experimental approaches can enhance genome annotation accuracy.

Techniques such as RNA sequencing (RNA-seq) enable the identification of expressed genes and alternative splicing events. By analyzing transcriptomic data, researchers can refine gene models and obtain insights into gene regulation.

## 18.5 APPLICATIONS OF GENOME SEQUENCING AND ANNOTATION

The applications of genome sequencing and annotation are vast and transformative, impacting various fields of research and industry.

### 18.5.1 Medical and Clinical Applications

Genome sequencing has revolutionized personalized medicine by enabling the identification of genetic variants associated with diseases. By analyzing an individual's genome, healthcare providers can tailor treatments based on genetic predispositions, optimizing therapeutic interventions. Furthermore, genome sequencing aids in understanding the genetic basis of complex diseases, facilitating the development of targeted therapies and precision medicine strategies.

### 18.5.2 Agricultural and Environmental Applications

In agriculture, genome sequencing is used to enhance crop traits, such as yield, disease resistance, and stress tolerance. By annotating the genomes of economically important crops, researchers can identify genes associated with desirable traits and apply marker-assisted selection in breeding programs. Additionally, environmental genomics utilizes sequencing to study microbial communities, assess biodiversity, and monitor ecosystem health.

### 18.5.3 Evolutionary and Comparative Genomics

Genome sequencing and annotation are fundamental to evolutionary studies, allowing researchers to compare genomic data across species. By analyzing genetic similarities and differences, scientists can infer evolutionary relationships and gain insights into the mechanisms of evolution. Comparative genomics contributes to our understanding of speciation, adaptation, and the conservation of genetic material across diverse organisms.

## 18.6 CHALLENGES IN GENOME SEQUENCING AND ANNOTATION

Despite the advancements in genome sequencing and annotation, several challenges persist that researchers must address.

### 18.6.1 Complexity of Genomes

Many organisms, especially eukaryotes, possess complex genomes with repetitive sequences, structural variations, and non-coding regions. This complexity poses challenges in accurately assembling and annotating genomes, requiring sophisticated algorithms and extensive computational resources.

### 18.6.2 Data Volume and Management

The sheer volume of data generated from high-throughput sequencing technologies presents significant challenges in data management and analysis. Efficient storage, processing, and sharing of genomic data are critical to facilitate collaboration and ensure reproducibility in research.

### 18.6.3 Accuracy and Validation

Ensuring the accuracy of genome annotations is paramount for reliable biological interpretations. Experimental validation of predicted gene functions and structures is essential to confirm computational predictions and enhance confidence in the annotated data.

## 18.7 CONCLUSION

In conclusion, genome sequencing and annotation are foundational components of modern genomics, providing critical insights into the genetic makeup of organisms. The evolution of sequencing technologies, coupled with advances in computational tools, has transformed the landscape of genomic research, enabling the exploration of complex biological questions.

As researchers continue to unlock the secrets of genomes, the applications of sequencing and annotation will expand, contributing to advancements in medicine, agriculture, and our understanding of the biological world. Addressing the challenges in these fields will ensure that the potential of genomic research is fully realized, paving the way for innovative solutions to global challenges.

# CHAPTER 19

# FUNDAMENTALS OF PROTEOMICS

Proteomics is a rapidly advancing field in molecular biology that focuses on the comprehensive study of proteins, particularly their structures, functions, interactions, and modifications. As proteins are essential players in virtually all biological processes, understa

nding their dynamics is crucial for elucidating cellular mechanisms, disease states, and potential therapeutic targets.

## 19.1 UNDERSTANDING PROTEOMICS

### 19.1.1 Definition of Proteomics

Proteomics is defined as the large-scale study of proteins, particularly with respect to their functions and structures. The term was coined in the mid-1990s and is derived from the combination of "protein" and "genome," reflecting its relationship with genomics. While genomics provides insights into the genetic blueprint of an organism, proteomics focuses on the actual products of gene expression—the proteins themselves.

### 19.1.2 Importance of Proteomics

Proteins are the workhorses of the cell, involved in virtually every biological function, including metabolism, signaling, transport, and immune responses. Understanding proteomics is essential for several reasons:

- **Biological Function:** By studying proteins, researchers can gain insights into how biological systems function, how cells communicate, and how metabolic pathways are regulated.
- **Disease Mechanisms:** Proteomics plays a critical role in understanding the molecular mechanisms underlying diseases such as cancer, neurodegenerative disorders, and infectious diseases. By identifying aberrant protein expressions or modifications, scientists can uncover potential biomarkers for diagnosis and treatment.
- **Drug Development:** Knowledge of protein interactions and functions can inform drug discovery efforts, leading to the identification of new therapeutic targets and the development of novel pharmaceuticals.

## 19.2 TECHNIQUES IN PROTEOMICS

Proteomics encompasses a variety of techniques that facilitate the identification, quantification, and characterization of proteins.

### 19.2.1 Protein Extraction and Separation

Before analyzing proteins, they must be extracted from biological samples and separated based on various properties. Common techniques include:

- **Gel Electrophoresis:** This technique separates proteins based on their size and charge. Sodium dodecyl sulfate-polyacrylamide gel electrophoresis (SDS-PAGE) is widely used for denaturing protein samples, allowing for the assessment of molecular weight and purity.
- **Isoelectric Focusing (IEF):** This method separates proteins based on their isoelectric point (pI), where proteins migrate in a pH gradient until they reach a point where their net charge is zero.

### 19.2.2 Mass Spectrometry (MS)

Mass spectrometry is a cornerstone technique in proteomics, enabling the identification and quantification of proteins with high sensitivity and specificity. The process involves three main steps:

- **Ionization:** Proteins are converted into charged ions, typically using methods like electrospray ionization (ESI) or matrix-assisted laser desorption/ionization (MALDI).
- **Mass Analysis:** The ions are analyzed based on their mass-to-charge ratios (m/z), allowing for the determination of their molecular weights.
- **Data Interpretation:** The mass spectra generated are analyzed to identify proteins and their sequences by comparing them to known protein databases.

### 19.2.3 Liquid Chromatography (LC)

Liquid chromatography, often coupled with mass spectrometry (LC-MS), is used to separate complex mixtures of proteins. Techniques like high-performance liquid chromatography (HPLC) and reversed-phase liquid chromatography are commonly employed to fractionate proteins based on their hydrophobicity, enabling better detection and analysis.

### 19.2.4 Protein Microarrays

Protein microarrays allow the simultaneous analysis of thousands of proteins in a high-throughput manner. These arrays consist of immobilized proteins on a solid surface, enabling researchers to study protein-protein interactions, post-translational modifications, and antibody specificity.

## 19.3 APPLICATIONS OF PROTEOMICS

The applications of proteomics span numerous fields, reflecting its significance in basic research and clinical practice.

### 19.3.1 Biomedical Research

In biomedical research, proteomics is employed to uncover the molecular mechanisms of diseases. By comparing the proteomes of healthy and diseased tissues, researchers can identify potential biomarkers for early diagnosis and therapeutic targets.

For instance, in cancer research, proteomic analyses have revealed altered expression patterns of oncogenes and tumor suppressor proteins, leading to insights into tumorigenesis.

### 19.3.2 Drug Development

Proteomics facilitates drug discovery by identifying novel protein targets for therapeutic intervention. By understanding protein interactions and signaling pathways, researchers can design small molecules or biologics that specifically modulate protein function, leading to more effective and targeted treatments.

### 19.3.3 Personalized Medicine

The integration of proteomics with genomics and metabolomics has paved the way for personalized medicine. By analyzing an individual's proteome, clinicians can tailor treatments based on the specific molecular characteristics of a patient's disease, improving therapeutic outcomes and minimizing side effects.

### 19.3.4 Agricultural and Environmental Applications

Proteomics is increasingly applied in agricultural research to enhance crop traits, such as disease resistance and stress tolerance. By studying the proteomes of plants, researchers can identify proteins involved in critical physiological processes, aiding in the development of improved crop varieties. Additionally, proteomics is used to monitor environmental changes and assess the impacts of pollutants on ecosystems by analyzing changes in protein expression in response to environmental stressors.

## 19.4 CHALLENGES IN PROTEOMICS

Despite the advancements in proteomics, several challenges remain that hinder the full realization of its potential.

### 19.4.1 Protein Complexity

Proteins exhibit a vast range of complexities, including variations in size, post-translational modifications (PTMs), and structural conformations. The presence of these complexities can complicate the identification and quantification of proteins, necessitating sophisticated techniques and tools for accurate analysis.

### 19.4.2 Data Management and Analysis

The high-throughput nature of proteomic studies generates massive amounts of data that require effective management and analysis. Bioinformatics tools and software are essential for processing and interpreting proteomic data, but the integration of data from multiple sources and formats poses significant challenges.

### 19.4.3 Standardization and Reproducibility

The lack of standardized protocols and methodologies in proteomics can lead to variability in results and hinder reproducibility across studies. Establishing standardized practices and guidelines for sample preparation, data acquisition, and analysis is crucial for ensuring consistency and reliability in proteomic research.

## 19.5 FUTURE DIRECTIONS IN PROTEOMICS

The field of proteomics is continually evolving, driven by technological advancements and the need for deeper biological insights.

### 19.5.1 Integration with Multi-Omics Approaches

Combining proteomics with genomics, transcriptomics, and metabolomics—collectively known as multi-omics approaches—can provide a more comprehensive understanding of biological systems. Integrating data from different omics layers can unveil complex regulatory networks and enhance our understanding of cellular processes.

### 19.5.2 Advancements in Mass Spectrometry

Emerging mass spectrometry techniques, such as top-down proteomics and single-cell proteomics, hold promise for advancing our knowledge of protein function and heterogeneity. These advancements can enable researchers to analyze intact proteins and explore cellular variability at the single-cell level, leading to new insights in biology and disease.

### 19.5.3 Clinical Applications

As proteomics continues to mature, its applications in clinical settings are likely to expand. The identification of novel biomarkers and therapeutic targets could facilitate the development of personalized treatment strategies and improve patient outcomes in various diseases.

## 19.6 CONCLUSION

In summary, proteomics is a dynamic and essential field of study that provides crucial insights into the complexities of protein biology. Through various techniques, researchers can analyze protein expression, structure, and function, paving the way for advancements in medicine, agriculture, and our understanding of biological systems.

Despite the challenges that remain, the future of proteomics holds great promise for transforming our approach to scientific discovery and innovation. As technologies evolve and become more refined, the potential applications of proteomics will continue to expand, significantly impacting health and disease management.

# CHAPTER 20

# FUNDAMENTALS OF METABOLOMICS

Metabolomics is an emerging field in the life sciences that focuses on the systematic study of metabolites within biological systems. Metabolites are small molecules that are produced during cellular metabolism and play vital roles in biochemical pathways. Understanding these compounds provides insights into the physiological state of an organism and its response to environmental changes, diseases, and therapies.

## 20.1 UNDERSTANDING METABOLOMICS

### 20.1.1 Definition of Metabolomics

Metabolomics is defined as the comprehensive analysis of metabolites in biological samples, such as tissues, cells, or bodily fluids, at a given time.

This field seeks to identify, quantify, and characterize the diverse array of metabolites present in a biological system. Unlike genomics or proteomics, which focus on genes and proteins, respectively, metabolomics provides a snapshot of the biochemical state of an organism by examining the end products of cellular processes.

### 20.1.2 Importance of Metabolomics

Metabolomics is crucial for several reasons:

- **Biochemical Insights:** By analyzing metabolites, researchers can gain valuable insights into the biochemical pathways and processes occurring within an organism. This information can help identify metabolic dysregulation in diseases, understand metabolic functions, and reveal new metabolic pathways.
- **Disease Diagnosis:** Metabolomics holds significant promise in the field of clinical diagnostics. Metabolite profiles can serve as biomarkers for early disease detection, monitoring disease progression, and evaluating therapeutic responses. This capability is particularly beneficial for complex diseases like cancer, diabetes, and cardiovascular conditions.
- **Pharmacometabolomics:** The study of how metabolism affects

drug efficacy and toxicity is known as pharmacometabolomics. By understanding individual metabolic profiles, researchers can personalize drug treatments, leading to improved therapeutic outcomes and reduced adverse effects.

## 20.2 TECHNIQUES IN METABOLOMICS

Metabolomics encompasses various techniques that facilitate the analysis of metabolites.

The primary methodologies can be grouped into two major categories: analytical techniques and data analysis tools.

### 20.2.1 Analytical Techniques

Analytical techniques are essential for the detection, identification, and quantification of metabolites in biological samples. Some of the most commonly used methods include:

- **Mass Spectrometry (MS):** Mass spectrometry is a key tool in metabolomics, offering high sensitivity and specificity for detecting a wide range of metabolites. MS can be coupled with various separation techniques, such as gas chromatography (GC-MS) and liquid chromatography (LC-MS), to enhance the identification and quantification of metabolites. The process involves ionizing the sample, analyzing the mass-to-charge ratios of the ions produced, and interpreting the resulting spectra.
- **Nuclear Magnetic Resonance (NMR) Spectroscopy:** NMR spectroscopy is a non-destructive analytical technique that provides detailed information about the structure and dynamics of metabolites. NMR allows for the identification of metabolites based on their unique chemical environments and can quantify metabolites directly from complex biological mixtures without extensive sample preparation.
- **Chromatography:** Chromatographic techniques, including gas chromatography (GC) and high-performance liquid chromatography (HPLC), are widely employed for separating metabolites based on their physicochemical properties. These methods can be used in conjunction with mass spectrometry or UV detection to enhance the analysis of complex metabolite mixtures.
- **Capillary Electrophoresis (CE):** CE is a technique that separates ions based on their size and charge using an electric field. It is particularly useful for analyzing small metabolites and ions, offering high resolution and rapid analysis times.

### 20.2.2 Data Analysis Tools

The vast amount of data generated in metabolomic studies necessitates the use of advanced data analysis tools and bioinformatics approaches. Key components include:

- **Statistical Analysis:** Statistical methods are crucial for interpreting metabolomic data. Techniques such as principal component analysis (PCA) and partial least squares-discriminant analysis (PLS-DA) help identify patterns and differences in metabolite profiles among various conditions or treatments.
- **Metabolite Identification Databases:** The identification of metabolites often requires comparison with known databases, such as METLIN and HMDB (Human Metabolome Database). These databases provide information on the chemical properties, biological functions, and mass spectra of metabolites, aiding in their identification.
- **Pathway Analysis Tools:** Tools like MetaboAnalyst and KEGG (Kyoto Encyclopedia of Genes and Genomes) facilitate the integration of metabolomic data with metabolic pathways. These resources help researchers understand the biological significance of observed metabolic changes and their implications for health and disease.

## 20.3 APPLICATIONS OF METABOLOMICS

The applications of metabolomics are vast and continue to expand across multiple disciplines.

### 20.3.1 Biomedical Research

In biomedical research, metabolomics is instrumental in understanding the metabolic changes associated with diseases. By comparing metabolite profiles from healthy and diseased tissues, researchers can identify potential biomarkers for early diagnosis and prognosis.

For instance, in cancer research, metabolomic studies have revealed alterations in metabolic pathways that could serve as diagnostic markers or therapeutic targets.

### 20.3.2 Nutritional Science

Metabolomics plays a significant role in nutritional science by providing insights into how dietary components affect metabolism and health. By analyzing metabolite profiles in response to different diets or nutritional interventions, researchers can identify biomarkers of dietary intake and assess the impact of nutrition on health outcomes. This information is valuable for developing personalized dietary recommendations.

### 20.3.3 Pharmacology

Metabolomics is increasingly used in pharmacology to understand drug metabolism, efficacy, and toxicity. By profiling metabolites before and after drug administration, researchers can assess how drugs are metabolized and identify potential side effects or adverse reactions. This information can lead to the development of safer and more effective pharmaceuticals.

### 20.3.4 Environmental Monitoring

Metabolomics is also applied in environmental studies to assess the impact of pollutants on living organisms. By analyzing the metabolite profiles of organisms exposed to environmental stressors, researchers can identify metabolic alterations that indicate exposure to harmful substances.

This application is essential for understanding ecosystem health and the effects of pollution on wildlife.

## 20.4 CHALLENGES IN METABOLOMICS

Despite its potential, metabolomics faces several challenges that researchers must address to advance the field.

### 20.4.1 Sample Complexity

Biological samples often contain a wide variety of metabolites, ranging from highly abundant compounds to low-abundance signaling molecules.

This complexity can make it difficult to detect and quantify all metabolites accurately. Additionally, the dynamic nature of metabolism means that metabolite levels can change rapidly in response to environmental or physiological changes.

### 20.4.2 Standardization and Reproducibility

The lack of standardized protocols for sample collection, preparation, and analysis can lead to variability in metabolomic studies. Establishing standardized practices is crucial for ensuring reproducibility and reliability in metabolomics research.

### 20.4.3 Data Interpretation

The interpretation of metabolomic data can be challenging due to the complexity of biological systems and the multitude of metabolites involved. Developing robust data analysis tools and frameworks is essential for extracting meaningful insights from metabolomic datasets.

## 20.5 FUTURE DIRECTIONS IN METABOLOMICS

The future of metabolomics holds great promise, driven by technological advancements and the increasing recognition of its importance in various fields.

### 20.5.1 Integration with Multi-Omics Approaches

Combining metabolomics with other omics disciplines, such as genomics and proteomics, can provide a more holistic understanding of biological systems. Integrating multi-omics data can reveal complex interactions between metabolites, genes, and proteins, leading to new insights into cellular function and disease mechanisms.

### 20.5.2 Advancements in Analytical Technologies

Ongoing advancements in analytical techniques, such as improvements in mass spectrometry resolution and sensitivity, will enhance the ability to detect low-abundance metabolites and increase the throughput of metabolomic analyses. These advancements will facilitate the analysis of more complex biological samples and expand the scope of metabolomic research.

### 20.5.3 Clinical Applications

As metabolomics continues to mature, its applications in clinical settings are expected to expand. The identification of novel biomarkers for disease diagnosis and prognosis, along with the development of personalized treatment strategies, will have a significant impact on healthcare and precision medicine.

## 20.6 CONCLUSION

In summary, metabolomics is a dynamic and rapidly evolving field that provides essential insights into the biochemical processes underlying health and disease. Through various analytical techniques and data analysis tools, researchers can analyze metabolites, uncover metabolic pathways, and identify biomarkers with significant clinical implications. While challenges remain, the future of metabolomics is bright, with the potential to transform our understanding of biology and enhance disease diagnosis and treatment. As technologies continue to advance, the integration of metabolomics with other omics approaches will further deepen our understanding of complex biological systems, ultimately contributing to improved health outcomes and personalized medicine.

# CHAPTER 21

# ENZYMES AND RECEPTORS

Enzymes and receptors are crucial biomolecules that play fundamental roles in biological processes. Enzymes function as biological catalysts that speed up chemical reactions in the body, while receptors are proteins that receive and transmit signals from outside the cell to elicit a response.

## 21.1 INTRODUCTION TO ENZYMES

### 21.1.1 Definition of Enzymes

Enzymes are specialized proteins that catalyze biochemical reactions by lowering the activation energy required for the reaction to proceed. They are highly specific, meaning each enzyme typically catalyzes a particular reaction or acts on a specific substrate. Enzymes are vital for a wide range of cellular processes, including metabolism, DNA replication, and cellular signaling.

### 21.1.2 Mechanism of Action

The mechanism by which enzymes operate involves several key steps. First, an enzyme binds to its substrate, forming an enzyme-substrate complex. This interaction stabilizes the transition state and allows the reaction to proceed more easily. The enzyme then catalyzes the conversion of substrate into product, which is subsequently released, leaving the enzyme free to catalyze additional reactions. The specificity of enzymes is determined by their active site, a unique region that precisely fits the substrate.

### 21.1.3 Factors Affecting Enzyme Activity

Enzyme activity can be influenced by various factors, including temperature, pH, substrate concentration, and the presence of inhibitors or activators. Each enzyme has an optimal temperature and pH range at which it functions best. Deviations from these conditions can lead to denaturation or reduced activity. Additionally, inhibitors can bind to enzymes and decrease their activity, while activators can enhance their catalytic efficiency.

## 21.2 INTRODUCTION TO RECEPTORS

### 21.2.1 Definition of Receptors

Receptors are proteins located on the cell membrane or within the cell that bind to specific signaling molecules, such as hormones, neurotransmitters, or drugs. This binding initiates a cellular response, allowing the cell to communicate with its environment. Receptors are essential for various physiological processes, including sensory perception, immune responses, and neurotransmission.

### 21.2.2 Types of Receptors

Receptors can be classified into several categories based on their structure and mechanism of action:

- **Membrane Receptors:** These receptors span the cell membrane and include G protein-coupled receptors (GPCRs), receptor tyrosine kinases (RTKs), and ion channel receptors. Membrane receptors are crucial for signal transduction, where the binding of a ligand triggers a cascade of intracellular events.
- **Intracellular Receptors:** Located within the cell, these receptors bind to ligands that can pass through the cell membrane, such as steroid hormones. Once activated, intracellular receptors can directly influence gene expression by acting as transcription factors.

### 21.2.3 Signal Transduction Mechanism

The binding of a ligand to a receptor initiates a series of events known as signal transduction.

This process often involves the activation of intracellular signaling pathways that result in a specific cellular response. For example, upon binding a ligand, a GPCR may activate G proteins, which in turn activate or inhibit downstream effectors, such as enzymes or ion channels. This amplification of the signal allows for a rapid and coordinated cellular response.

## 21.3 ENZYME-RECEPTOR INTERACTIONS

### 21.3.1 Functional Relationship

Enzymes and receptors are interconnected in various biological pathways. Many signaling cascades involve the activation of enzymes by receptors. For instance, upon binding of a hormone to its receptor, a receptor may activate an associated enzyme that catalyzes the production of second messengers, such as cyclic AMP (cAMP) or diacylglycerol.

These second messengers further propagate the signal within the cell and lead to a physiological response.

### 21.3.2 Enzyme Inhibition by Receptor Activation

In some cases, receptor activation can lead to the inhibition of specific enzymes. For example, the binding of a neurotransmitter to its receptor may initiate a signaling pathway that results in the phosphorylation and subsequent inhibition of an enzyme involved in neurotransmitter synthesis or degradation. This regulatory mechanism ensures precisc control over cellular processes and maintains homeostasis.

## 21.4 APPLICATIONS IN MEDICINE

### 21.4.1 Drug Design Targeting Enzymes and Receptors

The understanding of enzyme and receptor biology has profound implications for drug discovery and development. Many therapeutic agents are designed to interact specifically with enzymes or receptors to modulate their activity. For example, inhibitors that target specific enzymes are widely used in the treatment of diseases such as hypertension, cancer, and diabetes. Additionally, receptor agonists and antagonists are employed to influence signaling pathways in conditions like depression, anxiety, and pain.

### 21.4.2 Biologics and Monoclonal Antibodies

Biologics, including monoclonal antibodies, are engineered to target specific receptors or enzymes with high precision. These therapeutic proteins can either block or activate receptor functions, providing a powerful approach to treat various diseases. For instance, monoclonal antibodies targeting immune checkpoint receptors have revolutionized cancer therapy by enhancing the immune system's ability to recognize and eliminate tumor cells.

## 21.5 CHALLENGES AND FUTURE DIRECTIONS

### 21.5.1 Understanding Complex Interactions

Despite significant advancements in our understanding of enzymes and receptors, the complexity of their interactions remains a challenge. The dynamic nature of these proteins and their interactions with other cellular components complicates the prediction of their behavior in different biological contexts. Advanced techniques, such as high-throughput screening and structural biology, are being employed to unravel these complexities and enhance our understanding of enzyme-receptor interactions.

### 21.5.2 Personalized Medicine

The future of medicine increasingly relies on personalized approaches that consider individual variations in enzyme and receptor expression and function. Genetic differences can affect drug metabolism and receptor responsiveness, leading to variations in therapeutic outcomes. Personalized medicine aims to tailor treatments based on an individual's unique biochemical makeup, optimizing therapeutic efficacy and minimizing adverse effects.

## 21.6 CONCLUSION

Enzymes and receptors are fundamental components of biological systems, driving biochemical reactions and facilitating cellular communication. Their intricate interactions play critical roles in numerous physiological processes and are essential for maintaining homeostasis.

Understanding the mechanisms of enzyme action and receptor signaling has significant implications for medicine, particularly in drug design and therapeutic strategies. As research continues to advance, the potential for developing targeted therapies that leverage these biological molecules will enhance our ability to treat a wide range of diseases, ultimately improving patient outcomes and advancing the field of healthcare.

# CHAPTER 22

# DRUG DISCOVERY AND DEVELOPMENT

Drug discovery and development is a complex, multi-step process that transforms a novel idea into a viable therapeutic agent. This intricate journey involves numerous scientific disciplines and requires collaboration across various sectors, including academia, biotechnology, and pharmaceutical industries. The ultimate goal of drug discovery is to identify and develop new medications that effectively treat diseases while ensuring safety and efficacy for patients.

## 22.1 INTRODUCTION TO DRUG DISCOVERY

### 22.1.1 Definition of Drug Discovery

Drug discovery refers to the initial phase of the pharmaceutical development process that focuses on the identification of new compounds that can potentially be developed into drugs. This stage encompasses various scientific methods, including target identification, compound screening, and lead optimization. The process often begins with understanding the underlying biological mechanisms of a disease and identifying potential molecular targets for intervention.

### 22.1.2 Historical Context

Historically, drug discovery was a trial-and-error process often relying on traditional herbal remedies and serendipitous findings. However, the advent of modern science, particularly in the fields of biochemistry, molecular biology, and pharmacology, has significantly transformed the landscape. The introduction of high-throughput screening technologies and computer-aided drug design has enabled researchers to efficiently evaluate large libraries of compounds and predict their interactions with biological targets.

## 22.2 STAGES OF DRUG DISCOVERY

### 22.2.1 Target Identification and Validation

The first critical step in drug discovery is identifying a biological target, such as a protein, enzyme, or receptor, associated with a specific disease. This process involves extensive research to understand the disease mechanism and identify pathways that can be modulated therapeutically. Once a potential target is identified, validation studies are conducted to confirm its role in the disease and to ensure that manipulating the target will yield a therapeutic effect.

### 22.2.2 Hit Identification

After establishing a validated target, the next phase is to identify "hits" or lead compounds that can interact with the target. High-throughput screening (HTS) methods are often employed during this stage, allowing researchers to rapidly test thousands of compounds for their ability to bind to the target and induce a desired biological effect. Hits are initially assessed for their efficacy and specificity before moving forward in the development process.

### 22.2.3 Lead Optimization

Once promising hits are identified, the focus shifts to lead optimization, where chemical modifications are made to enhance the compound's efficacy, selectivity, and pharmacokinetic properties. This iterative process often involves synthesizing and testing numerous analogs to identify the most effective and safest candidate. In this stage, researchers aim to improve the drug's absorption, distribution, metabolism, and excretion (ADME) characteristics while minimizing toxicity.

## 22.3 PRECLINICAL DEVELOPMENT

### 22.3.1 In Vitro and In Vivo Studies

Before progressing to human clinical trials, lead candidates undergo extensive preclinical testing to evaluate their safety and efficacy. This stage typically involves in vitro studies using cell cultures to assess the compound's biological activity and potential toxicity. Following promising in vitro results, in vivo studies in animal models are conducted to gain insights into pharmacokinetics, pharmacodynamics, and potential side effects.

### 22.3.2 Regulatory Considerations

Preclinical studies are critical for obtaining regulatory approval to conduct human trials. Regulatory agencies, such as the U.S. Food and Drug Administration (FDA), require comprehensive documentation of preclinical findings before granting Investigational New Drug (IND) status. This status allows researchers to begin clinical trials in humans while ensuring that the potential benefits of the drug outweigh any risks.

## 22.4 CLINICAL DEVELOPMENT

### 22.4.1 Phases of Clinical Trials

Clinical development comprises several phases, each designed to evaluate different aspects of the drug's safety and efficacy.

- **Phase I Trials:** The primary goal of Phase I trials is to assess the safety, tolerability, and pharmacokinetics of the drug in a small group of healthy volunteers. Researchers closely monitor participants for adverse reactions and determine the appropriate dosage range.
- **Phase II Trials:** Phase II trials involve a larger group of patients and focus on evaluating the drug's efficacy and side effects in individuals with the targeted disease. These studies help establish optimal dosing regimens and provide initial data on the drug's therapeutic potential.
- **Phase III Trials:** Phase III trials are pivotal studies that involve thousands of participants across multiple sites. They are designed to confirm the drug's efficacy, monitor adverse effects, and compare it to existing standard treatments. Successful Phase III trials are essential for obtaining regulatory approval for market access.

### 22.4.2 Regulatory Approval

After completing successful Phase III trials, the next step is to submit a New Drug Application (NDA) to regulatory authorities. This comprehensive application includes all data from preclinical and clinical studies, as well as information on the drug's manufacturing processes and labeling. Regulatory agencies evaluate the NDA to ensure that the drug is safe and effective for public use.

## 22.5 POST-MARKETING SURVEILLANCE

### 22.5.1 Phase IV Trials

Once a drug is approved and on the market, post-marketing surveillance, also known as Phase IV trials, continues to monitor the drug's long-term safety and efficacy. These studies help identify rare side effects and assess the drug's performance in a broader population. Regulatory agencies may require additional studies to further evaluate the drug under real-world conditions.

### 22.5.2 Adverse Event Reporting

Pharmaceutical companies are responsible for monitoring and reporting adverse events associated with their drugs. Healthcare professionals and patients are encouraged to report any unexpected side effects to regulatory authorities, ensuring ongoing safety assessments and prompt action if necessary.

## 22.6 CHALLENGES IN DRUG DISCOVERY AND DEVELOPMENT

### 22.6.1 High Failure Rates

The drug discovery process is notoriously challenging, with high attrition rates at every stage. It is estimated that only a small fraction of compounds identified in early discovery ever make it to market. Factors contributing to this high failure rate include lack of efficacy, unforeseen side effects, and challenges in scaling up production.

### 22.6.2 Cost and Time

Drug discovery and development is a time-consuming and costly endeavor, often taking over a decade and billions of dollars to bring a new drug to market. These financial and temporal constraints pose significant challenges, particularly for small biotechnology companies that may lack the resources of larger pharmaceutical firms.

### 22.6.3 Regulatory Hurdles

Navigating the regulatory landscape is another significant challenge in drug development. Regulatory agencies impose stringent requirements to ensure the safety and efficacy of new drugs, which can prolong the development timeline. Furthermore, the need for comprehensive data and evidence can lead to additional costs and complexities.

## 22.7 FUTURE TRENDS IN DRUG DISCOVERY

### 22.7.1 Precision Medicine

The future of drug discovery is increasingly leaning towards precision medicine, which aims to tailor treatments to individual patient characteristics, including genetics, lifestyle, and environment.

Advances in genomics and biotechnology are enabling the development of personalized therapies that target specific molecular profiles, improving treatment efficacy and reducing adverse effects.

### 22.7.2 Artificial Intelligence and Machine Learning

The integration of artificial intelligence (AI) and machine learning (ML) into drug discovery processes holds great promise for enhancing efficiency and effectiveness. These technologies can analyze vast datasets to identify potential drug candidates, predict their interactions, and optimize drug design, significantly accelerating the discovery process.

### 22.7.3 Drug Repurposing

Another emerging trend is drug repurposing, which involves finding new therapeutic uses for existing drugs. This approach can drastically reduce development timelines and costs, as the safety profiles of repurposed drugs are often well-established. Drug repurposing has gained attention during public health emergencies, such as the COVID-19 pandemic, where existing medications were rapidly evaluated for efficacy against the novel virus.

## 22.8 CONCLUSION

Drug discovery and development is a complex and multi-faceted process that requires the integration of various scientific disciplines, innovative technologies, and collaborative efforts. From the initial identification of targets to the final regulatory approval of new therapies, each stage presents unique challenges and opportunities.

As the landscape of drug development continues to evolve, embracing advances in precision medicine, artificial intelligence, and drug repurposing will be crucial for overcoming existing hurdles and improving patient outcomes. Ultimately, the goal remains the same: to deliver safe and effective medications that enhance health and quality of life for individuals around the globe.

# BIBLIOGRAPHY


**Biochemical Engineering** Atkinson, B., & Floudas, C.A. (2008). Wiley.

**Biochemistry and Molecular Biology** Ahern, D. (2016). Cengage Learning.

**Biochemistry and Molecular Biology Education** O'Connor, C.B., & Wylie, A. (2017). Wiley.

**Biochemistry** Berg, J.M. (2017). W.H. Freeman.

**Biochemistry** Berg, J.M., & Tymoczko, J.L. (2007). W.H. Freeman and Company.

**Biochemistry** Nelson, D.L., & Cox, M.M. (2013). W.H. Freeman and Company.

**Biochemistry of Lipids: Metabolism and Membrane** F. D. (2018). Academic Press.

**Biochemistry** Stryer, L. (2019). W.H. Freeman and Company.

**Biochemistry** Voet, D., & Voet, J.G. (2010). John Wiley & Sons.

**Biochemistry: A Short Course** Rodwell, V.W., et al. (2015). Cengage Learning.

**Biochemistry: A Student Companion** C. A. G. (2014). Academic Press.

**Biochemistry: The Molecular Basis of Life** Ghosh, D. (2019). Oxford University Press.

**Bioinformatics: Principles and Applications** D. R. (2018). Springer.

**Bioinformatics: Sequence and Genome Analysis** Mount, D.W. (2004). Cold Spring Harbor Laboratory Press.

**Biophysical Chemistry** C. R. (2016). Wiley.

**Cell and Molecular Biology: Concepts and Experiments** Karp, G. (2016). Wiley.

**Cell Biology** Alberts, B., & Johnson, A. (2018). Garland Science.

**Cell Biology** Thomas D. Pollard, William C. Earnshaw, & Jennifer Lippincott-Schwartz. (2017). Elsevier.

**Enzymes: Biochemistry, Biotechnology, Clinical Chemistry** McDonald, J. (2018). Wiley.

**Essentials of Glycobiology** Varki, A., Cummings, R.D., Esko, J.D., & Freeze, H.H. (2017). Cold Spring Harbor Laboratory Press.

**Fundamentals of Biochemistry: Life at the Molecular Level** Donald Voet, Judith G. Voet, & Charlotte W. Pratt. (2016). Wiley.

**Genetics: A Conceptual Approach** Hartl, D.L., & Jones, E.W. (2011). Jones & Bartlett Learning.

**Introduction to Protein Structure** Branden, C. & Tooze, J. (1999). Garland Science.

**Lehninger Principles of Biochemistry** Nelson, D.L., & Cox, M.M. (2017). W.H. Freeman and Company.

**Molecular Biology of the Cell** Alberts, B., Johnson, A., Lewis, J., Raff, M., Roberts, K., & Walter, P. (2014). Garland Science.

**Molecular Biology of the Gene** Watson, J.D., & Baker, T.A. (2017). Cold Spring Harbor Laboratory Press.

**Molecular Biology Techniques: A Classroom Laboratory Manual** Karp, G. (2017). Wiley.

**Molecular Biology: A Comprehensive Introduction** C. G. K. (2016). Academic Press.

**Molecular Biology: Principles and Practice** Cox, M.M., & Dunlap, K. (2015). W.H. Freeman and Company.

**Molecular Genetics of Bacteria** Snyder, L. & Champness, J. (2013). ASM Press.

**Principles of Biochemistry** Berg, J.M., & Tymoczko, J.L. (2012). W.H. Freeman and Company.

**Principles of Biochemistry** Horton, H.R., et al. (2016). Pearson.

**Principles of Biochemistry** Rodwell, V.W. (2014). Cengage Learning.

# INDEX

## W



## X



## Y



## Z

www.ingramcontent.com/pod-product-compliance
Lightning Source LLC
LaVergne TN
LVHW062118200926
848484LV00007B/1708